T0255662

Practical C++20 Financial Programming

Problem Solving for Quantitative Finance, Financial Engineering, Business, and Economics

Second Edition

Carlos Oliveira

Apress®

Practical C++20 Financial Programming: Problem Solving for Quantitative Finance, Financial Engineering, Business, and Economics

Carlos Oliveira
Seattle, WA, USA

ISBN-13 (pbk): 978-1-4842-6833-9   ISBN-13 (electronic): 978-1-4842-6834-6
https://doi.org/10.1007/978-1-4842-6834-6

Managing Director, Apress Media LLC: Welmoed Spahr
Acquisitions Editor: Steve Anglin
Development Editor: Matthew Moodie
Coordinating Editor: Mark Powers

Cover designed by eStudioCalamar

Cover image by Fly-D on Unsplash (www.unsplash.com)

Distributed to the book trade worldwide by Apress Media, LLC, 1 New York Plaza, New York, NY 10004, U.S.A. Phone 1-800-SPRINGER, fax (201) 348-4505, e-mail orders-ny@springer-sbm.com, or visit www. springeronline.com. Apress Media, LLC is a California LLC and the sole member (owner) is Springer Science + Business Media Finance Inc (SSBM Finance Inc). SSBM Finance Inc is a **Delaware** corporation.

For information on translations, please e-mail booktranslations@springernature.com; for reprint, paperback, or audio rights, please e-mail bookpermissions@springernature.com.

Apress titles may be purchased in bulk for academic, corporate, or promotional use. eBook versions and licenses are also available for most titles. For more information, reference our Print and eBook Bulk Sales web page at http://www.apress.com/bulk-sales.

Any source code or other supplementary material referenced by the author in this book is available to readers on GitHub via the book's product page, located at www.apress.com/9781484268339 and on coliveira.net/. For more detailed information, please visit http://www.apress.com/source-code.

Printed on acid-free paper

To my family, my real source of inspiration.

Table of Contents

About the Author

Carlos Oliveira works in the area of optimization and quantitative finance, with more than 15 years of experience in creating scientific and financial models in C++. During his career, Carlos has developed several large-scale applications for financial companies such as Bloomberg L.P. and Incapital LLC. Carlos obtained a PhD in Operations Research and Systems Engineering from the University of Florida, an MSc in Computer Science from UFC (Brazil), and a BSc in Computer Science from UECE (Brazil). He also performs academic research in the field of combinatorial optimization, with applications in diverse areas such as finance, telecommunications, computational biology, transportation, and logistics. Carlos has written more than 30 academic papers on optimization and authored four books, including *Options and Derivatives Programming in C++20: Algorithms and Programming Techniques for the Financial Industry*, Second Edition (Apress, 2020).

Introduction

This is a hands-on book for programmers who want to learn about how C++20 is used in the financial industry. The book concentrates on the parts of the language that are more frequently used to write financial software, including the STL (standard template library), templates, and support for numerical libraries. I also describe many of the important problems in financial engineering that are part of the day-to-day work of financial programmers and quantitative analysts in investment banks and hedge funds.

The book provides how-to examples that cover all the major tools and concepts used to build working solutions for financial applications. Each chapter teaches readers how to use advanced C++ concepts as well as the basic building libraries used by modern C++ developers, such as the STL and Boost. I discuss how to create correct and efficient applications, leveraging knowledge of object-oriented and template-based programming. I assume only a basic knowledge of C and C++ and build on these concepts to explain techniques already mastered by developers who are familiar with modern C++.

In the process of writing this book, I was concerned with providing a great value for readers who are trying to use their programming knowledge to become proficient in the style of programming used in financial institutions such as banks, hedge funds, and other companies in the financial industry. However, I have introduced the topics covered in the book in a logical and structured way, so that even novice programmers will be able to absorb the most important topics and competencies necessary to develop financial applications in C++.

An important feature of the book is its focus on specific themes and practical solutions for financial problems. While the emphasis is not on the theoretical aspects of finance, I do discuss topics such as numerical algorithms, integration techniques, and differential equations for derivative valuation. Moreover, the reader will gain a good understanding of how to model such problems using modern C++ concepts.

The financial literature for programmers typically has a large number of books written from an academic standpoint, with most of the time spent on the discussion of mathematics concepts behind algorithms, rather than the software engineering challenges that developers need to overcome. Therefore, in this book, I decided to focus on working solutions for common programming problems, in the form of code examples, offering readers much more value for their reading efforts.

Audience

This book is intended for readers who already have a working knowledge of programming in C, C++, or another mainstream language. These are usually professionals or advanced students in computer science, engineering, physics, and mathematics, who have an interest in learning C++20 financial programming either for personal improvement or for professional reasons. The book is also directed at practitioners of C++ programming in financial institutions, who would use the book as a ready-to-use reference for common development problems.

By reading this book, you will learn how to use modern C++20 techniques to implement practical applications. Being a multi-paradigm language, C++ is used slightly differently in each application area. Therefore, skills that are valuable for developing desktop applications are not necessarily the same as those used to write high-performance software. A large part of major high-performance financial applications are written in C++, which means that programmers who want to enter this lucrative market need to acquire a working knowledge of a few specific and relevant parts of the language. This book therefore would be an excellent choice for developers who want to advance their knowledge effectively while learning one of the most sought-after and marketable skill sets for modern applications and high-performance software development.

Content Overview

Here is a brief overview of the contents of each chapter.

Chapter 1—The Fixed Income Market: The fixed income market is a large part of the financial engineering industry, and it presents unique computational challenges for its practitioners. C++ programming is widely used in this area, offering the ability to compute rates and cash flow variations with incredible speed, as readers will learn in this chapter. I present C++ coding examples that can be used in the solution of some of the most common problems occurring in fixed income markets. I include C++ algorithms for topics such as (1) interest rate calculation, (2) present value computation, (3) cash flows, and (4) valuation of bonds.

Chapter 2—The Equities Market: Equity markets are multifaceted and offer a great variety of investment vehicles. As a result, the number and complexity of computational techniques used for financial analysis of equity markets continue to grow. In this chapter, I present C++ examples for a few selected problems occurring in the equities markets and their derivatives. I cover programming topics such as the following: (1) moving average computation, (2) calculating volatility, (3) computing instrument correlation, and (3) calculating fundamental indicators.

Chapter 3—C++ Programming Techniques in Finance: The C++ language was created as an extension of C, which means that most programs written in C are also valid C++ programs. However, good C++ programs need to make use of high-level features made available by the language to control program complexity. This is especially important for financial applications, where we want to create fast and expressive applications. In this chapter, I explore fundamental techniques that financial C++ programmers use to write better code with less effort, including (1) class templates, (2) auto pointers, (3) shared pointers, (4) resource acquisition is initialization (RAII), (5) automatic type detection, (6) exception handling, and (7) operator overloading.

Chapter 4—Common Libraries for Financial Applications: Modern coding in C++ uses libraries that simplify the creation of fast, standard-conforming classes. The STL offers a set of generic, standard containers that can be used in almost any situation. Knowing how to use the STL well is one of the main skills necessary for effective C++ programming. Another common set of classes is contained in the Boost libraries, which are usually the basis for the next version of the C++ standard. Readers will learn about topics such as (1) STL containers, (2) STL algorithms, (3) boost libraries, and (4) date and time handling.

Chapter 5—Designing Numerical Classes: At the heart of financial applications is a set of well-designed numerical classes. This chapter tells you how to create numerical classes that will perform efficiently when used in production code. You will also see examples in C++ that show how to integrate with existing numerical classes and algorithms. You will learn how to (1) implement a matrix class, (2) perform calculations at compilation time with templates, (3) represent ratios with C++ templates, and (4) generate statistical data.

Chapter 6—Plotting Financial Data: A common activity in financial programming is the generation of data that needs to be visualized by traders or other financial stakeholders. Most of the time, the data needs to be plotted in the form of a chart for easy visualization. I give a few examples that show how to plot data in C++ programs using common libraries. You will learn about topics such as (1) using Gnuplot to plot data, (2) designing a class to create Gnuplot charts, and (3) plotting from a GUI (graphical user interface) application using Qt.

Chapter 7—Linear Algebra: Linear algebra (LA) techniques are used throughout the area of financial engineering. Therefore, it is important to understand how the traditional methods of LA can be applied in C++. With this goal in mind, I present a few examples that show how to use some of the most common LA algorithms. In this chapter, you will also learn about (1) integrating existing LA libraries into your code, (2) basic LA operations, (3) the BLAS (basic linear algebra subprograms) library, and (4) calculating the determinant of a matrix with BLAS.

Chapter 8—Interpolation: Interpolation is a commonly used technique that finds a mathematical function approximating a set of points. Fast interpolation is the secret for high-performance algorithms in several areas of financial engineering. This chapter will show you programming samples that cover a few of the most common interpolation methods, with efficient implementation in C++. The main techniques discussed in this chapter are (1) linear interpolation and (2) polynomial interpolation.

Chapter 9—Calculating Roots of Equations: Equations are one of the building blocks of algorithms in financial engineering, and it is important to be able to calculate equation roots efficiently. In this chapter, you will find algorithms for different methods of calculating equation roots, along with explanations of how they work and when they should be used. Topics include (1) the bisection method, (2) the secant method, and (3) Newton's method.

Chapter 10—Numerical Integration: Function integration is a common part of many financial algorithms. However, it is hard to solve certain classes of equations exactly, and numerical methods need to be employed in such cases. In this chapter, you will see examples of C++ code that can be readily applied to common integration problems. I also discuss the performance and the accuracy of such methods. The programming examples in this chapter cover topics such as (1) the midpoint method, (2) the trapezoid method, and (3) Simpson's method.

Chapter 11—Solving ODEs and PDEs: Differential equations are at the heart of many techniques used in the analysis of equity markets. There are several processes for solving and analyzing ordinary differential equations (ODE) and partial differential equations (PDE) that can be implemented in C++. In this chapter, I present programming examples that cover aspects of ODEs and PDE modeling and application in C++. Topics covered include the following: (1) solving ODEs, (2) using the Runge-Kutta method, and (3) solving the Black-Scholes equation.

Chapter 12—Optimization: Optimization refers to a set of techniques used to find the minimum or maximum of a function. Optimization strategies are used in several areas of financial engineering. In this chapter, I discuss programming techniques that can be used to implement common aspects of optimization algorithms. I provide a concise explanation of some techniques and how they are typically implemented in C++20. You will learn about (1) modeling optimization problems, (2) interfacing with linear programming (LP) solvers, (3) solving two-dimensional LP problems, and (4) mixed integer–programming models.

Chapter 13—Asset and Portfolio Optimization: Portfolio managers have to face the issue of balancing a portfolio for optimal performance, depending on their predefined portfolio goals. Optimization-based techniques have been developed to deal with some of the most common portfolio construction problems. In this chapter, we consider algorithms for portfolio optimization using C++. We consider how to design such optimization code in order to get results that are as fast and as accurate as possible. Topics include (1) creating a portfolio model, (2) performing resource allocation, and (3) using linear techniques for portfolio optimization.

Chapter 14—Monte Carlo Methods: Among other programming techniques used in equity markets analysis, Monte Carlo simulation has a special place due to its wide applicability and easy implementation. These methods can be used to forecast prices or to validate buying strategies, for example. In this chapter, I provide programming examples that can be used as part of simulation-based algorithms, with topics such as (1) random number generation, (2) optimization through Monte Carlo methods, and (3) simulation models for price forecasting.

Chapter 15—Extending Financial Libraries: C++ is a complete language that can be used to develop the most complex software. However, it is sometimes beneficial to combine C++ libraries with scripting languages that can simplify the creation of prototypes and other noncritical applications. In this chapter, I show you how to use the solutions and algorithms discussed in the text as external libraries for scripting languages that are commonly employed in the financial industry. In particular, you will learn how to (1) extend C++ with Python and (2) extend C++ with Lua scripts.

Chapter 16—Using C++ Code with R and Maxima: Financial algorithms in C++ can be used not only as part of executable code but also as part of other modeling and development environments. In this chapter, I show you how to integrate financial libraries into two well-known simulation and modeling environments for financial analysis: R and Maxima. You will see how it is possible to create loadable modules for these environments, incorporating complex C++ algorithms in a way that they are ready to use from scripts written in R and Maxima.

Chapter 17—Multithreading: Financial applications have very stringent performance requirements. A common way to improve response time is to use concurrency and parallel programming techniques, such as multithreading. C++ can be used to write very responsive multithreaded applications, and in this chapter, I explore algorithms for creating and managing threads, with applications to financial problems. I also cover the important topic of data access synchronization. Topics include (1) creating threads, (2) protecting shared memory, (3) synchronization techniques, and (4) threads using the standard library.

Appendix A—C++20 Features: C++ is an evolving language, and in the last few years, we have seen a renewed effort to bring much-needed updates. The latest efforts are the C++17 and C++20 standards, and major C++ compilers are incorporating these features at a fast pace. In the appendix, I cover examples that show how some of these features can improve your code and simplify the development of new programs and libraries. You will learn about new features such as (1) auto variables, (2) closures, (3) rvalues, (4) const expressions, and (5) initializer lists.

Introduction to the Second Edition

In this second edition of the book, the examples and the text have been revised to conform to the latest C++ standard, C++20. While much of our examples continue to compile and work properly in the new standard, we felt the need to present new C++ features that will make it easier to develop financial applications.

For example, the appendix now presents some new features only available in C++20. We also explain how to use threads in the standard library, among other improvements. All examples have been tested to make sure that we conform to the latest standard.

Compiling the Code Samples

The examples given in this book have all been tested on Windows using the MingW gcc compiler and on Mac OS X using the Xcode 12 IDE. You should be able to build the code, however, using any standards-compliant C++ compiler that implements the C++20 standard. For example, gcc is available on Linux and other platforms, and Microsoft Visual Studio will also work on Windows.

If you use Mac OS X and don't have Xcode installed in your computer, you can download it for free from Apple's developer website at http://developer.apple.com. The code can also be compiled from the command line, as it is explained in each chapter.

If you instead want to use MingW on Windows, you can download it from the website www.mingw.org.

Once MingW is installed, start the command prompt from the MingW program group in the start menu. Then, you can type gcc to check that the compiler is properly installed.

To download the complete set of examples, visit the web page for this book at http://coliveira.net, or navigate to apress.com/9781484268339 and click the **Download Source Code** button.

CHAPTER 1

The Fixed Income Market

The fixed income market is a large part of the financial industry, and it presents unique challenges and opportunities for its practitioners. A large amount of the money managed by pension funds and other institutional funds is allocated to fixed income investments. Because fixed income has a predictable income stream, conservative money managers view it as a safer investment option when compared to stocks and more exotic derivatives. As a result, traditional institutions commit a lot of time and effort to the fixed income industry.

As software engineers, our main goal when working in the fixed income market is to define computational strategies and solve problems so that our clients can be successful. C++ is a language that is uniquely poised to the solution of problems in this industry. This is due to its flexibility and high performance on standard computational platforms. Moreover, C++ is a highly portable language that can be used in a variety of computer systems.

As a result of the advantages just mentioned, C++ programing has been widely used in this area of finance, and it is one of the preferred languages used in banks, hedge funds, pension funds, and other large institutions that have to deal with fixed income as one of their main investment vehicles. Programmers who work with C++ have over the years developed software that offers useful capabilities for fixed income analysis, such as computing prevailing interest rates and determining cash flow valuations. All of these features need to execute with incredible speed, with the help of some of the techniques explored in later sections of this book. Due to its new standard, C++20, the language is nowadays even more capable of satisfying the strict requirements demanded by the financial industry.

© Carlos Oliveira 2021
C. Oliveira, *Practical C++20 Financial Programming*, https://doi.org/10.1007/978-1-4842-6834-6_1

In this chapter, I provide a quick introduction to this area of finance and show you a few C++ coding examples that can be used in the solution of some of the most common programming problems occurring in fixed income markets. These coding examples include the solution to problems involving

- Simple interest rate calculation

- Compound interest rate calculation

- Cash flow modeling

- Determination of the present value of cash flows

- Modeling and valuation of bonds

In the remainder of this chapter, I will also show you why C++20 may be the ideal language to deal with programming problems occurring in the financial investment industry and in particular how to solve problems in fixed income investing. Then, I will provide a general introduction to the issues occurring in fixed income investments and an overview of how the fixed income market works. Then, I will start with a few programming examples that explore the concepts discussed in the previous sections.

Fixed Income Overview

We start our discussion with a general overview of fixed income instruments. While this is not a book on finance or economics, it is still important to have a few concepts in place. My general goal is to describe how to use these concepts in the solution of the practical computational problems that we discuss in the latter part of this chapter.

In a fixed income investment, a contractually defined exchange occurs between two parties. Both parties agree to exchange cash flows that are assigned based on interest rates and the time of cash exchanges. Fixed income investments are very diverse, but they include the following well-known types of investments vehicles:

- Money market funds: These are short-term investments that offer a small rate of return but at the same time provide easy availability of funds at your own convenience. Money market funds have a very short-term horizon, and they only pay returns that are close to the spot rate practiced by banks. Since money market funds have a small return that is hard to predict over a long period, they are used mostly for their liquidity.

- Bonds: This is a major category of fixed income applications. Bonds pay a predetermined interest rate for a well-defined period of time. They are issued by a variety of institutions, including companies and all levels of government. The American government, for example, issues treasury bonds, which are one of the main investment vehicles used throughout the world.

- Certificates of deposit: These are fixed income investments issued by banks to their retail customers. They are simple investments that pay a fixed interest rate for a predefined period, usually between 1 and 5 years. They are used mainly for the convenience of small investors who lack access to more sophisticated fixed income markets and want to invest from their own checking or savings account.

The main reason for investors to enter the fixed income market is to take advantage of a relatively safe investment opportunity, where the returns are known and predictable. Compared to the stock market, fixed income investments have the advantage of being easier to analyze. This is true because, for equity investments, for example, it is practically impossible to determine how much money a company will make in a few years from now. With a fixed income investment such as a bond, however, you have a contract that guarantees the return on the investment for a specified period of time.

Clearly, there are also risks in such fixed income investments. A well-known risk is that of the default of the institution issuing the bond, for example. In that case, investors may lose a part of the, or the whole, investment. The second big risk, which is frequently overlooked by investors, is that the rate of return will not be able to cope with inflation during the period of the investment. For example, if the rate of return is 6% a year but inflation is around 4%, then your real rate of return is just 2% (and that is the return before taxes).

This all shows that analyzing fixed income investments is not as easy as it initially sounds. It is not just a matter of finding the institution paying the largest interest rate and putting all your money on its bonds. This is one of the reasons why money managers need reliable software that can be used to decide which is best among myriad fixed income investments. Just as the stock market presents thousands of possibilities that need to be carefully analyzed, the fixed income industry has a huge number of available choices. One of the big tasks for software developers is to create systems that can easily track these investments and help in choosing the right options for long-term investors.

Note   Fixed income investments have risks that are hard to measure because they depend on the future economic environment. Sound fixed income investments need to take into consideration the several risks involved. High-quality C++ software for fixed income may help investors to take into consideration some of these external factors.

Here are some of the most important concepts about fixed income investments used through this chapter.

- Interest rate: The return of investment in percentage points for a given period (usually 1 year). Fixed income investments will have a well-defined interest rate that is determined as a contractual obligation.

- Principal: The amount of the original fixed income loan or investment. This is the value over which the interest rate is calculated in the case of a fixed income investment such as a bond.

- Compound interest: Interest that is accrued over time and added to the principal as regular interest payments are made at each period. The amount of compound interest is regulated by the interval between interest payments.

- Continuous compounding: As the number of periods increase, the effect of compound interest becomes more pronounced. For example, compound interest paid at the end of every month will produce more than at a yearly payment schedule. In theory, this compounding process could happen in a continuous schedule, and the resulting compound interest can be calculated using a simple formula, which I explain later in this chapter.

- Present value: When a set of scheduled cash flows and an interest rate are defined, it is possible to calculate the present value of those cash flows. This is done using the contractual interest rate to determine the discounted value of each future cash flow and adding together all these values. The present value is a very powerful tool to compare two cash flow streams.

Using these simple concepts, it is possible to analyze very complex investments. You will learn how to use these concepts in some of the coding examples contained later in this chapter.

Why Use C++

C++ is a language that has been used with great success in all kinds of financial applications. It is the number one language used by Wall Street firms to create fast, high-performance code that can be employed to implement efficient algorithms for financial engineering.

While C++ is already a mature language with more than 30 years of history, and other programming languages have appeared since then with high-level features that are easier to use, C++ still holds the place as the standard language for high-performance computation. Large financial institutions such as banks, hedge funds, and pension funds rely daily on C++ to solve their most complex computational problems for the following reasons:

- Performance: The most obvious reason why C++ is used is its performance. Due to the fact that C++ has little runtime overhead compared to other high-level languages, it is possible to use it to write very fast software. Not only is C++ fast enough by default, but it also allows expert C++ programmers to explore many additional low-level techniques for code optimization, which are not available to programmers using languages such as Java and Python.

- Standards compliance: C++ is a standard language, developed over the years by an international group of experts with the goal of providing high-level features such as object-oriented programming (OOP) without the overhead that is normally associated with them. As a result of the standardization effort, C++ is available on all kinds of platforms, ranging from microcontrollers to the largest servers. This means that you can run your algorithms unmodified between platforms. This is an obvious advantage for financial algorithms, since this kind of software can be easily ported to faster architectures over the years to take advantage of improvements in new hardware and software design.

- Existing libraries: C++ offers an almost unparalleled set of libraries for numerical and financial programming. Each topic we discuss in this book has several libraries available that can save time and effort.

- Multi-paradigm language: Developers designed the C++ language from the beginning to support multiple programming paradigms, so programmers don't need to change the essence of an algorithm to fit into a particular paradigm. For example, although OOP is supported, the language does not mandate the use of OOP. In this way, programmers are free to use the most expressive technique for the desired application.

- High-level features: Although C++ allows programmers to achieve high performance by targeting low-level features of their hardware, good programmers can still use several high-level features that make C++ a truly modern language. For example, C++ was one of the first languages to embrace the concept of OOP, which is without question the most common paradigm for modern software design. C++ has also pioneered other features such as exceptions and template-based containers. More recently, C++ incorporated even more high-level features by means of the new C++11 standard of the language. Automated type detection, lambda expressions, and user-defined literals are just a few of the new features that have become available to application developers since the new standard was approved.

For the reasons stated previously, programmers have trusted C++ as the main vehicle for implementing high-performance financial algorithms. In this book, we explore code examples that make use of these computational advantages.

Like any other tool, C++ also has its share of problems. One of the themes in learning C++ programming is to avoid dangerous practices that can lead to bugs and unsafe programs. Most of the techniques you will see in the next chapters embrace the use of modern libraries, which not only simplify the process of creating C++ programs but also allow you to create software that is well designed and fault tolerant. Using the standard library, which includes the STL (standard templates library), is the best way to use C++ safely.

You will also learn how to use the high-quality libraries that have been made available through the boost project. The boost libraries have been designed from the ground up to use modern C++ concepts in a way that simplifies the creation of new software. The boost libraries are the result of the work of some of the greatest experts in C++ programming, including people involved in the C++ standard committee itself. In fact, many of the libraries shipped with boost have become part of the standard library. Therefore, using boost libraries, you will be getting early access to some features that will be included in future versions of the language.

Calculating Simple Interest Rates

To start, I will show you how to solve a very simple problem in fixed income analysis, as a way to introduce some of the features of C++ class design that we use throughout this book.

Problem

Interest rates determine how much a financial institution is going to pay in exchange for holding a cash deposit over a period of time. Calculate the future value of a deposit given the interest rate and the initial value of the deposit, assuming a single period of deposit.

Solution

You just need to use the mathematic equation for simple interest rate calculation, which is given by the expression

$$V = P \left(1 + R \right)$$

In this formula, V is the future value after a single period, and P is the present value of the deposit. With this formula, you can calculate the interest rate for a single period.

How It Works

The `IntRateCalculator` class, defined in Listing 1-1, determines the calculation of single-period interest rates.

Listing 1-1. The IntRateCalculator Class

```
class IntRateCalculator {
public:
    IntRateCalculator(double rate);
    IntRateCalculator(const IntRateCalculator &v);
    IntRateCalculator &operator =(const IntRateCalculator &v);
    ~IntRateCalculator();

    double singlePeriod(double value);
private:
    double m_rate;
};
```

First, we define a new class that becomes responsible for the calculation. A fundamental principle of object-oriented design is to have responsibilities unified under very well-defined interfaces. You should embrace this principle when creating C++ classes, since it will simplify maintenance and avoid costly mistakes. Even if you need to write additional code using this strategy, the increased organization pays off in the long run.

In the definition of the `IntRateCalculator` class, we define a constructor, a destructor, a copy constructor, and the assignment operator. These are methods that, if you don't define them yourself, will be added to the class by the compiler. It is useful to create your own versions of such member functions, however, because in this way, you can be sure that you are getting the desired behavior, instead of what the compiler writers think is the right choice.

Note   You should create classes that specify the four basic member functions automatically defined by the C++ compiler. In this way, you can avoid costly mistakes by having the created objects use a well-defined life cycle. Failing to provide such member functions can result in classes that don't respond correctly to such basic operations as assignment (defined by the assignment operator) and copy construction. If your class is supposed to be the base for other classes, you should also make the destructor virtual, so that the derived classes can properly release the resources they use. This way, the runtime system can properly detect the polymorphic type of the object and call the right destructor.

The compiler automatically adds the following member functions, unless you specify otherwise in the class declaration:

- The default constructor: The default constructor is automatically added, allowing an object to be created using the new keyword, even if the class writer didn't include it. A default constructor is one that has no arguments. It is not included automatically, however, if the class declaration contains another constructor that requires arguments. For example, in our IntRateCalculator class, the constructor receives one parameter, the interest rate. Therefore, the default constructor is not automatically included, which means that to create an object of the IntRateCalculator class, the programmer needs to specify a valid interest rate argument.

- The copy constructor: The copy constructor allows you to create copies of an existing object of the same class. It is included by default only if there are no other constructors in the class definition. In our case, we need to supply a copy constructor, to guarantee that it is possible to create copies of existing objects. Copy constructors become important when objects need to be added to containers, particularly the containers provided in the STL, such as vectors, maps, and multimaps.

- The destructor: A destructor defines how the resources used by a particular object will be freed once the object is destroyed. A proper constructor is required to avoid memory leaks and other undesirable resource leaks in an object. In the IntRateCalculator class, there are no internal or external resources that need to be freed, but it is still better to define this explicitly.

- The moving constructor: A moving constructor provides the operations used when the C++ moving semantic is required.

- The assignment operator: This member function is used when an assignment operation occurs between two objects of the same class. Defining this type, you can specify how the contents of an object are transferred from one object to the next: that can be done either by value or by reference. Other details of the copy, such as reference counters, for example, can also be established in the assignment operator.

9

The singlePeriod member function encapsulates the operation that returns the future value of a deposit after a single period. Depending on the structure of the loan or the input parameters, this can refer to 1 month or 1 year of interest. The signature of the member function is

```
double singlePeriod(double value);
```

This simple version of the code uses the double type (instead of float) for extra precision. In the next chapters, we will discuss how to deal with precision issues that are inherent to floating point numbers.

The IntRateCalculator class contains a single member variable, m_rate, which stores the current interest rate. In this way, it is not necessary to input the interest rate every time the singlePeriod member function is called. Therefore, to create a new instance of IntRateCalculator, you need to provide the interest rate as a parameter to the constructor.

The header file, IntRateCalculator.h, defines the singlePeriod member function as inline (see Listing 1-2).

```
inline double IntRateCalculator::singlePeriod(double value)
{
    double f = value * ( 1 + this->m_rate );
    return f;
}
```

The keyword inline is used here to suggest that the member function be directly embedded in the code that calls it. What this means is that there is no penalty for calling this function, since the function call will be removed from the executed code, and the content of the method will be directly substituted. Think of this as a way of achieving the same performance of a macro, with all the compiler support of calling a function. In high-performance C++ code, it is common to see member functions defined as inline, in order to achieve even higher performance than equivalent member function calls. This kind of flexibility is one of the features that separate C++ from other languages, where it would be much more difficult to achieve similar performance.

Complete Code

Listing 1-2. IntRateCalculator.h

```
//
//  IntRateCalculator.h

#ifndef __FinancialSamples__IntRateCalculator__
#define __FinancialSamples__IntRateCalculator__

#include <iostream>

class IntRateCalculator {
public:
    IntRateCalculator(double rate);
    IntRateCalculator(const IntRateCalculator &v);
    IntRateCalculator &operator =(const IntRateCalculator &v);
    ~IntRateCalculator();

    double singlePeriod(double value);
private:
    double m_rate;
};

inline double IntRateCalculator::singlePeriod(double value)
{
    double f = value * ( 1 + this->m_rate );
    return f;
}

#endif /* defined(__FinancialSamples__IntRateCalculator__) */

//
//  IntRateCalculator.cpp

#include "IntRateCalculator.h"

IntRateCalculator::IntRateCalculator(double rate)
: m_rate(rate)
```

```
{

}

IntRateCalculator::~IntRateCalculator()
{

}

IntRateCalculator::IntRateCalculator(const IntRateCalculator &v)
: m_rate(v.m_rate)
{

}

IntRateCalculator &IntRateCalculator::operator=(const IntRateCalculator &v)
{
    if (&v != this)
    {
        this->m_rate = v.m_rate;
    }
    return *this;
}
//
//  main.cpp

#include "IntRateCalculator.h"

#include <iostream>

// the main function receives parameters passed to the program
int main(int argc, const char * argv[])
{
    if (argc != 3)
    {
        std::cout << "usage: progName <interest rate> <value> "
        << std::endl;
        return 1;
    }
```

```
    double rate = atof(argv[1]);
    double value = atof(argv[2]);

    IntRateCalculator irCalculator(rate);
    double res = irCalculator.singlePeriod(value);
    std::cout << " result is " << res << std::endl;
    return 0;
}
```

Sample Use

First, you need to compile the code using your favorite C++ compiler. For example, using the makefile provided in a UNIX platform, you could just use the make command, with the following results:

```
$ make
gcc -c IntRateCalculator.cpp
gcc -c main.cpp
gcc -o intrate IntRateCalculator.o main.o
```

You can now run this program by passing a given interest rate and initial value. For example, you could type the following:

```
./intrate 0.08 10000
 result is 10800
```

This shows that the future value of an investment of $10,000 at an 8% interest rate is $10,800 after a single period.

Compound Interest

You can use simple interest rates to analyze single-period cash flows. However, most financial operations, such as loans, have multiple periods. For this purpose, you need to consider compound interest.

Problem

Calculate the compound interest accumulated by a given principal value after the passage of N time periods.

Solution

The solution uses a new C++ class that encapsulates the concept of compound interest. With this class, it becomes easy to answer the proposed question using two member functions. The first function, `multiplePeriod`, returns the future value of a fixed income investment after a given number of periods, as passed in the function parameter.

As mentioned previously, interest can be calculated either as a discrete or a continuous compounding process. For discrete compounding, we assume that interest is paid only at regular intervals, as defined by the investment vehicle. The compounding happens as interest is added to the original principal.

The formula for discrete compounded interest rate is

$$V = P(1+R)^{N}$$

where P is the present value, V is the future value, R is the interest rate, and N is the number of periods. The interest rate is the value passed as a parameter to the class constructor and stored as a member variable. The number of periods N is passed as the second parameter to the `multiplePeriod` method.

For continuous compounding calculation, you need to use a separate method, `continuousCompounding`. In this case, we assume that compounding doesn't happen in discrete steps but that the payments are made continuously over time. This is a possible way to determine the future value of a financial application (or at least an upper bound for the desired future value).

The formula for the calculation of continuous interest rate compounding is

$$V = Pe^{RN}$$

Here, V is the desired future value, P is the present value, R is the interest rate during the period, and N is the number of periods. For example, to find the future value of continuously compounded interest after 2 years at 8% interest per year, you should use the value of the previous equation with parameters R = 0.08 and N = 2.

How It Works

The two member functions, `multiplePeriod` and `continuousCompounding`, calculate the given formulas using the mathematical functions `pow` and `exp` from the standard C++ library. These two functions implement a fast way to calculate the power function and the exponential function, respectively.

To use any mathematical function from the standard library, you should first include the header file `cmath`. Table 1-1 provides a short list of mathematical functions made available from that header file.

Table 1-1. *Some of the Mathematical Functions in the Standard Library*

Function	Corresponding mathematical operations
exp	Exponential function (natural base)
pow	Power function
log	Natural logarithm function
log10	Logarithm function on decimal base
sqrt	Square root function
Sin	Sine function
cos	Cosine function
tan	Tangent function
acos	Arc cosine function (inverse of cosine)
asin	Arc sine function (inverse of sine)
atan	Arc tangent function (inverse of tangent)
ceil	Ceiling function (smallest integer higher than parameter)
floor	Floor function (largest integer lower than parameter)
fabs	Absolute value for float numbers

The mathematical functions provided by the standard library should be used whenever possible, instead of custom versions, for the following reasons:

- Compatibility: Using functions from the standard library guarantees that they will be available in any compiler that implements it.

- Performance: Functions in the standard library are implemented as part of the package sold by compiler vendors. The code of these mathematical functions is generally optimized for the particular architecture, which usually results in much better performance.

Complete Code

The code in Listing 1-3 shows the implementation for class CompoundIntRateCalculator, divided into a header file and an implementation file. I also present a sample main function that shows how to use the class.

Listing 1-3. CompoundIntRateCalculator.h

```
//
// CompoundIntRateCalculator.h

#ifndef __FinancialSamples__CompoundIntRateCalculator__
#define __FinancialSamples__CompoundIntRateCalculator__

class CompoundIntRateCalculator {
public:
    CompoundIntRateCalculator(double rate);
    CompoundIntRateCalculator(const CompoundIntRateCalculator &v);
    CompoundIntRateCalculator &operator =(const CompoundIntRateCalculator &v);
    ~CompoundIntRateCalculator();

    double multiplePeriod(double value, int numPeriods);
    double continuousCompounding(double value, int numPeriods);
 private:
    double m_rate;
};
```

```cpp
#endif /* defined(__FinancialSamples__CompoundIntRateCalculator__) */

//
//  CompoundIntRateCalculator.cpp

#include "CompoundIntRateCalculator.h"

#include <cmath>

CompoundIntRateCalculator::CompoundIntRateCalculator(double rate)
: m_rate(rate)
{

}

CompoundIntRateCalculator::~CompoundIntRateCalculator()
{

}

CompoundIntRateCalculator::CompoundIntRateCalculator(const
CompoundIntRateCalculator &v)
: m_rate(v.m_rate)
{

}

CompoundIntRateCalculator &CompoundIntRateCalculator::operator =(const
CompoundIntRateCalculator &v)
{
    if (this != &v)
    {
        this->m_rate = v.m_rate;
    }
    return *this;
}
```

```
double CompoundIntRateCalculator::multiplePeriod(double value, int
numPeriods)
{
    double f = value * pow(1 + m_rate, numPeriods);
    return f;
}

double CompoundIntRateCalculator::continuousCompounding(double value, int
numPeriods)
{
    double f = value * exp(m_rate * numPeriods);
    return f;
}

//
//  main.cpp

#include "CompoundIntRateCalculator.h"

#include <iostream>

// the main function receives parameters passed to the program
int main(int argc, const char * argv[])
{
    if (argc != 4)
    {
        std::cout << "usage: progName <interest rate> <present value> <num
        periods>" << std::endl;
        return 1;
    }

    double rate = atof(argv[1]);
    double value = atof(argv[2]);
    int num_periods = atoi(argv[3]);

    CompoundIntRateCalculator cIRCalc(rate);
    double res = cIRCalc.multiplePeriod(value, num_periods);

    double contRes = cIRCalc.continuousCompounding(value, num_periods);
```

```
std::cout << " future value for multiple period compounding is " << res
<< std::endl;
std::cout << " future value for continuous compounding is " << contRes
<< std::endl;

return 0;
}
```

Sample Use

The code in Listing 1-3 can be compiled into an executable and run from the command line. The program expects three arguments: the interest rate, the present value of the investment, and the number of periods of compounding.

The following is an example of its use:

```
$ ./compound 0.05 1000 4
 future value for multiple period compounding is 1215.51
 future value for continuous compounding is 1221.4
```

As expected, the value returned by continuous compounding is slightly higher than the value achieved by discrete compounding.

Modeling Cash Flows

A more general way of thinking about fixed income investments is to look at the flow of cash exchanged between the two involved parties. A cash flow is a sequence of payments, scheduled during a specified period of time. It is clear that the value of the cash flows between two entities should be equal in some way. In this section, you will learn how to determine if a set of cash flows is equivalent.

Problem

Calculate the present value of two cash flows and determine if they are equivalent.

Solution

Cash flows are the basic tool for comparing two or more fixed income investments. A cash flow establishes the sequence of cash transfers between two interested parties. The traditional way to denote these cash exchanges is by using positive and negative values.

For example, consider a common loan, where a customer requests a quantity at a given interest rate. The customer will make a sequence of cash payments during the lifetime of the loan. At the end of the transaction, the payments made by both parties should be equivalent.

The equivalence is established using the concept of *present value*. The present value of a payment in the future needs to be discounted by the interest rate that would be applied to that same value. In other words, discounting is the inverse concept to compounding.

Calculating Present Value

A general principle of investing is that money in your pocket today is more valuable than the same money received in the future. This general principle can be quantified using the knowledge of value compounding based on interest rates. The present value of a fixed income investment is the value that corresponds to the sum of cash flows taking place in the future, after their corresponding interest has been considered and discounted.

The formula for present value (PV) of a future payment is determined by

$$PV = FV / (1 + R)^N$$

In this equation, PV is the desired present value, FV is the future value that we want to discount, R is the interest rate, and N is the number of periods between the present value and the future value.

As you see, the formula for PV is the inverse of the calculation of compound interest rate. This clearly shows that we are just using a similar process to determine a present value when starting from a known future value.

Calculating Present Value in C++

Formulas for calculating PV can be found in any financial engineering book. For a C++ programmer, however, the main interest in this topic is centered on how to perform PV calculations with high performance. The standard procedure is to denote values paid by the two parties using positive and negative signs. For example, we can denote an initial loan as a negative number and each payment of the loan as a positive number. Using this approach, for a cash flow from two parties to be equivalent, the present value of all cash transfers needs to add to zero.

This is the method used by the CashFlowCalculator class, which is presented next. Here is the class definition.

```cpp
class CashFlowCalculator {
public:
    // constructors

    void addCashPayment(double value, int timePeriod);
    double presentValue();
private:
    std::vector<double> m_cashPayments;
    std::vector<int> m_timePeriods;
    double m_rate;
    double presentValue(double futureValue, int timePeriod);
};
```

The addCashPayment method is used to add new payments to the desired cash flow. The arguments are the value of the payment, and the second is the time period when this payment occurs. The value is positive or negative depending on the originator of the payment, as previously discussed. The data is stored on two vectors, m_cashPayments and m_timePeriods, using the STL vector template.

The presentValue method in this class is used to compute the PV or the whole cash flow stored in the current object. This is done with the determination of the PV for each cash exchange as stored in the m_cashPayments vector and finally adding these values to the total variable.

```
double CashFlowCalculator::presentValue()
{
    double total = 0;
    for (int i=0; i<m_cashPayments.size(); ++i)
    {
        total += presentValue(m_cashPayments[i], m_timePeriods[i]);
    }
    return total;
}
```

The auxiliary member function presentValue(double, int) is used to calculate the PV for a single payment. It is defined using the foregoing formula.

```
double CashFlowCalculator::presentValue(double futureValue, int timePeriod)
{
    double pValue = futureValue / pow(1+m_rate, timePeriod);
    std::cout << " value " << pValue << std::endl;
    return pValue;
}
```

Using STL Containers

The code in the CashFlowCalculator class is made simpler by the use of vector containers. The std::vector<> template is used in modern C++ applications to store ordered sequences of elements that require random access. Unlike traditional C and C++ arrays, which decay to pointers when passed as arguments to a function, a vector is an object that maintains its properties, such as size, during the whole time the vector is used. A vector also knows how to clean up after itself, avoiding memory leaks that are so common in old-style C++ applications.

To use a vector in a C++ application, you need to declare the object by passing the element type as a parameter to the vector template. Therefore, std::vector<int> will create a vector of int elements. The vector template class has member functions that can be used to manipulate and retrieve information about the elements.

- `size`: Returns the number of elements stored in the vector object.

- `push_back`: Copies the object passed as a parameter and stores it at the end of the vector. If necessary, additional memory is allocated for the new element, which can take O(n).

- `pop_back`: Removes the last element from the vector and undoes the changes made by push_back (except for memory that is not released).

- `operator[]`: Provides access to the contents of the vector, using syntax similar to the access of traditional C++ arrays.

The vector template is just one among other STL containers that are available for C++ developers. The complete list changes as new templates are added to the standard library, but Table 1-2 lists the most used containers.

Table 1-2. *Common Containers Provided by the STL*

Container	Description
vector	Ordered collection of elements with constant random access time
queue	Container where elements are added at the end and removed from the front position
map	Associative container that connects keys to their associated element
multimap	Associative container that connects keys to a set of associated elements
list	A linked list of elements, which provides constant time inclusion/exclusion at any position
stack	A specialized container that allows only addition and removal of the last element (the top of the stack)

Complete Code

Listing 1-4 presents the code for the class `CashFlowCalculator`. The code is divided into a header file and an implementation file. You can see how to use the code in the example shown in the section "Running the Code."

Listing 1-4. CashFlowCalculator.h

```
//
// CashFlowCalculator.h

#ifndef __FinancialSamples__CashFlowCalculator__
#define __FinancialSamples__CashFlowCalculator__

#include <vector>

class CashFlowCalculator {
public:
    CashFlowCalculator(double rate);
    CashFlowCalculator(const CashFlowCalculator &v);
    CashFlowCalculator &operator =(const CashFlowCalculator &v);
    ~CashFlowCalculator();

    void addCashPayment(double value, int timePeriod);
    double presentValue();
private:
    std::vector<double> m_cashPayments;
    std::vector<int> m_timePeriods;
    double m_rate;
    double presentValue(double futureValue, int timePeriod);
};

#endif /* defined(__FinancialSamples__CashFlowCalculator__) */

//
// CashFlowCalculator.cpp

#include "CashFlowCalculator.h"

#include <cmath>
#include <iostream>

CashFlowCalculator::CashFlowCalculator(double rate)
: m_rate(rate)
{

}
```

```cpp
CashFlowCalculator::CashFlowCalculator(const CashFlowCalculator &v)
: m_rate(v.m_rate)
{

}

CashFlowCalculator::~CashFlowCalculator()
{

}

CashFlowCalculator &CashFlowCalculator::operator =(const CashFlowCalculator
&v)
{
    if (this != &v)
    {
        this->m_cashPayments = v.m_cashPayments;
        this->m_timePeriods = v.m_timePeriods;
        this->m_rate = v.m_rate;
    }
    return *this;
}

void CashFlowCalculator::addCashPayment(double value, int timePeriod)
{
    m_cashPayments.push_back(value);
    m_timePeriods.push_back(timePeriod);
}

double CashFlowCalculator::presentValue(double futureValue, int timePeriod)
{
    double pValue = futureValue / pow(1+m_rate, timePeriod);
    std::cout << " value " << pValue << std::endl;
    return pValue;
}
```

```cpp
double CashFlowCalculator::presentValue()
{
    double total = 0;
    for (int i=0; i<m_cashPayments.size(); ++i)
    {
        total += presentValue(m_cashPayments[i], m_timePeriods[i]);
    }
    return total;
}

//
//  main.cpp

#include "CashFlowCalculator.h"

#include <iostream>

// the main function receives parameters passed to the program
int main(int argc, const char * argv[])
{
    if (argc != 2)
    {
        std::cout << "usage: progName <interest rate>" << std::endl;
        return 1;
    }

    double rate = atof(argv[1]);

    CashFlowCalculator cfc(rate);
    do {
        int period;
        std::cin >> period;
        if (period == -1) {
            break;
        }
        double value;
        std::cin >> value;
        cfc.addCashPayment(value, period);
```

```
} while (1);

double result = cfc.presentValue();
std::cout << " The present value is " << result << std::endl;

return 0;
}
```

Running the Code

The program can be compiled using a standards-compliant C++ compiler such as GCC on Linux or Mac OS X. The resulting program can be executed in the following way:

```
./presentValue 0.08
1 200
2 300
3 500
4 -1000
-1
 value 190.476
 value 272.109
 value 431.919
 value -822.702
The present value is 71.8014
```

The first few lines display the input for the program. The command line argument (in this case 0.08) is the desired interest rate—it is used as the parameter to the class constructor. The following lines are a sequence of the time periods and payment values. The last line of the sequence is marked using the number -1. When that number is read, the program stops reading the input and starts to calculate the PV of the given cash transfers, in the order in which they were received.

The last few lines display the output of the program. The code prints the PV for each component of the cash flow. Finally, it prints the PV of the whole sequence of payments. To use this program to validate a common fixed income instrument, such as a loan, you should input each pair of time period–payment value. At the end of the calculation, the PV should add to zero (or close to zero, due to possible numerical inaccuracies).

Modeling Bonds

Bonds are a very common type of fixed income instrument. They are used by large corporations and governments all over the world to attract cash investments that will be repaid in the long term. In exchange, they offer the guaranteed payment of a periodic coupon. Most bonds mature (are paid off) in a time period between 5 and 30 years.

Problem

Create a C++ class to model a bond instrument and determine its annual interest rate.

Solution

Bonds are structured in such a way that the investor deposits the principal value at the beginning of the term of the bond. Frequently, the principal is repaid in its entirety at maturity. Between the period between the initial investment and its maturity, investors are paid a constant value, also called the coupon value, which determines the interest rate paid by the bond.

For example, consider a 30-year, $100,000 bond investment in company XYZ, with an annual coupon of $5,000. This translates into a fixed income investment that pays a 5% interest on the principal. Company XYZ has the right to use the principal during the specified period of time, and the total value of the principal is returned to the investor in 30 years at maturity.

To model this kind of investment using C++, you can create a class that contains the needed information, such as principal value, coupon value, and maturity period. The class has the following declaration:

```
class BondCalculator {
public:
    BondCalculator(const std::string institution, int numPeriods, double
    principal, double couponValue);
    BondCalculator(const BondCalculator &v);
    BondCalculator &operator =(const BondCalculator &v);
    ~BondCalculator();

    double interestRate();
```

```
private:
    std::string m_institution;
    double m_principal;
    double m_coupon;
    int m_numPeriods;
};
```

This class has member variables that store the name of the institution that originates the bond (known as the issuer), the principal invested, the coupon amount, and the number of periods (usually defined in years). The class can be used to record information about bond investments as part of an application that tracks such fixed income investments. The interestRate method can be used to return the internal rate of returned implied by the coupon.

Complete Code

Listing 1-5 shows a complete listing for class BondCalculator. The code is split into a header file and an implementation file. You can also check a sample usage contained in the main function.

Listing 1-5. BondCalculator.h

```
//
// BondCalculator.h

#ifndef __FinancialSamples__BondCalculator__
#define __FinancialSamples__BondCalculator__

class BondCalculator {
public:
    BondCalculator(const std::string institution, int numPeriods, double
    principal, double couponValue);
    BondCalculator(const BondCalculator &v);
    BondCalculator &operator =(const BondCalculator &v);
    ~BondCalculator();

    double interestRate();
```

```
private:
    std::string m_institution;
    double m_principal;
    double m_coupon;
    int m_numPeriods;
};

#endif /* defined(__FinancialSamples__BondCalculator__) */

//
//   BondCalculator.cpp

#include "BondCalculator.h"

BondCalculator::BondCalculator(const std::string institution, int numPeriods,
                               double principal, double couponValue)
: m_institution(institution),
  m_numPeriods(numPeriods),
  m_principal(principal),
  m_coupon(couponValue)
{

}

BondCalculator::BondCalculator(const BondCalculator &v)
: m_institution(institution),
  m_numPeriods(v.m_numPeriods),
  m_principal(v.m_principal),
  m_coupon(v.m_coupon)
{

}

BondCalculator::~BondCalculator()
{

}
```

```cpp
BondCalculator &BondCalculator::operator =(const BondCalculator &v)
{
    if (this != &v)
    {
        this->m_institution = v.m_institution;
        this->m_principal = v.m_principal;
        this->m_numPeriods = v.m_numPeriods;
        this->m_coupon = v.m_coupon;
    }
    return *this;
}

double BondCalculator::interestRate()
{
    return m_coupon / m_principal;
}

// the main function receives parameters passed to the program
int main(int argc, const char * argv[])
{
    if (argc != 4)
    {
        std::cout << "usage: progName <institution> <principal> <coupon>
        <num periods>"
                    << std::endl;
        return 1;
    }

    std::string issuer = argv[1];
    double principal = atof(argv[2]);
    double coupon = atof(argv[3]);
    int num_periods = atoi(argv[4]);

    BondCalculator bc(issuer, principal, coupon, num_periods);
    std::cout << "reading information for bond issued by " << issuer <<
    std::endl;
```

```
    std::cout << " the internal rate of return is " << bc.interestRate() <<
    std::endl;
    return 0;
}
```

Running the Code

The code can be compiled using a standards-compliant C++ compiler. It has been tested on Linux and Mac OS X. You can run the program using the following command at your preferred shell:

$ ./bondCalculator XYZ 100000 5000 20
```
reading information for bond issued by XYZ
 the internal rate of return is 0.5
```

The first line in bold is the command that you need to execute. The parameters are the name of the issuer institution, the total principal invested in the bond, the value of the periodic coupon, and the number of time periods for this investment.

The output of the program displays the rate of return calculated from the coupon value. The class BondCalculator can now be used in a larger application to store information about this type of fixed income investment.

Further Reference

This chapter provides an introduction to the general topic of fixed income investments. While we are mostly concerned about the C++ programming issues involved in this area, there are several books that can help you get a greater understanding of the financial engineering techniques that were introduced here.

The following books are just suggestions that you can explore to achieve a better understanding of the world of fixed income investments.

- *Investment Science* by David Luenberger (Oxford University Press, 1998): This is an undergraduate-level book that describes the basic theory of investment. Most of the book explains the fundamentals of fixed income investments, including algorithms for the most common problems.

- *Investments* by Zvi Bodie, Alex Kane, and Alan Marcus (McGraw-Hill/Irwin, 2004): This is a standard textbook on investment theory that explains, among other topics, the ideas behind fixed income investments.

- *Mathematics for Finance* by Marek Carpinski and Tomasz Zastawniak (Springer, 2011): This book is more for the mathematically inclined. It not only explains the basics of fixed income investments but also gives a lot of mathematical methods that are useful in their analysis.

Conclusion

In this chapter, I introduced the topic of fixed income investments and how they can be modeled and analyzed using C++ code. The first part of the chapter explains the general concepts behind fixed income investments. These investments are used as a relatively safe way to maintain and generate wealth, as compared to the equity and derivatives market.

I have also explained why C++, especially in its current standard C++20, is the ideal programming language to create computational solutions for the problems in this area of finance. Due to its performance characteristics and high-level programming support, C++ provides the best balance between expressiveness and raw speed. As a result, C++ is the de facto standard for the development of core applications in the finance field, especially in applications that deal with fixed income data.

The first example introduced a basic class that can be used to calculate simple interest rates. It introduces not only the concept of interest rate calculation methods but also the typical way such solutions are designed and coded in modern C++.

The second example introduced the concept of interest rate compounding, both in discrete and continuous intervals. You learned there how to create a C++ class to calculate this type of interest rate using standard C++ library functions. I presented a summary of such mathematical functions and how they are used in C++ programs.

The third example in this chapter explored the important concept of cash flows and their corresponding PV. The calculation of PV is central to the comparison of two or more fixed income investments. Using the inverse of the formulas for interest rate, you can determine the real value of a given set of cash flows in the present. You learned how to solve this type of problem using a new C++ class.

Finally, this chapter explains how bonds are used in financial applications and presented a class to model these investments. In future chapters, you will learn more about the computational challenges of using these financial vehicles as part of an investment portfolio.

In the next chapter, I will introduce another large part of the financial investment landscape: the equities market. You will see a few programming techniques that can be useful in these markets, along with an introduction to other important concepts that we explore in the later part of this book.

CHAPTER 2

The Equities Market

Owning shares of company profits is one of the most common ways to invest and generate wealth. A large number of people who have made a fortune have achieved it by creating or buying an equity stake in a successful corporation. This is the reason the equity market is so popular among all kinds of investors. Moreover, the stock market is so vast that it provides opportunities for everyone willing to participate: from small investors to large hedge funds, you will find an investment style for each kind of participant.

The equities market is also an exciting area for software engineers, since it provides so many opportunities to apply computational techniques, which can be implemented in C++. Software engineers are also great allies to market analysts and investors in general, helping in vital activities such as modeling market data and devising algorithms needed to make fast and accurate trading decisions.

Due to their large size, equity markets are multifaceted and offer a huge variety of investment vehicles. From small cap stocks to blue chips, ETFs (exchange-traded funds), equity and index options, and other derivatives, there are a great number of opportunities for employing investment algorithms, in order to get an edge in the market. As a result, there is also great incentive (from banks and other investment institutions) to apply high-speed C++ programming techniques to solve such problems.

In this chapter, we present C++ code for a few selected problems occurring in the equities markets and their derivatives. We will consider financial programming topics such as the following:

- Calculating simple moving averages

- Computing exponential moving averages

- Calculating volatility

- Computing correlation of equity instruments

- Modeling and calculating fundamental indicators

© Carlos Oliveira 2021
C. Oliveira, *Practical C++20 Financial Programming*, https://doi.org/10.1007/978-1-4842-6834-6_2

Equities Market Concepts

Equity markets exist to expedite the trading of equity-based investments. The goal of an equity investment is to allocate money directly or indirectly to company stock, which gives buyers a certain share of ownership in a company. The idea behind this investment is to profit from the growth of the institution represented by that particular investment vehicle. For example, buying shares of IBM stock gives ownership of a small part of the company, along with the future profits associated with that ownership.

Direct stock ownership is the simplest example of an equity investment. Anyone with a brokerage account can buy shares in public companies, that is, companies that have put their shares for sale in the public market. Using their particular trading accounts or retirement accounts, individual investors have the ability to invest in any one of the thousands of publicly traded companies in the US and international markets.

However, directly controlling a company stock is not the only (or even the easiest) way to participate in the stock market. There are nowadays a plethora of products that offer alternative ways to invest in equity. This includes mutual funds, ETFs, index funds, options, and other more exotic derivatives. How to select the right instrument from such a large array of tradable issues is one of the many problems faced by money managers and individual investors.

Market Participants

The equities market is composed of many participants. They have different goals and interests; however, they work continuously to maintain market prices while trying to profit from them.

Large institutions form a sizable portion of the equities market landscape. These big, sell-side investment institutions (such as investment banks and exchanges) are viewed as the backbone of the market. Therefore, they are also commonly referred to as market makers. These large companies are buying and selling great volumes of equity investment vehicles (such as stocks) daily, with the goal of having small profits in each operation. More recently, high-frequency trading was added to this picture, resulting in increased volume and speed in market transactions.

The following is a quick list of the most common players in the equities market:

- Mutual funds: These funds receive investments from retail investors and institutions and make investments in areas of the market that they believe will have larger than usual investment returns. Mutual funds are mostly limited to buying stocks and ETFs, so their performance is limited when the market is in a downtrend.

- Hedge funds: Hedge funds use more advanced techniques, such as shorting stocks and buying options and futures on risky investments not available to common investors, so they are limited to wealthier investors and some kinds of institutions that can cope with the increased risk.

- Investment banks: These institutions are actively working on the market composition. For example, they act in bringing to the market new issues (also known as IPOs) that will be traded by other investors. They are also allowed to trade for themselves and other large clients.

- High-frequency trading funds: These funds use high-performance computational techniques to provide instant liquidity to the markets while making small profits in a large number of transactions.

- Brokerage companies: These companies work directly with individual investors providing the ability to buy or sell stocks, ETFs, mutual funds, and options for a small or even no commission per transaction. Their services are made available through the Internet on several platforms such as desktops, web browsers, and mobile devices.

- Pension funds: These are institutions that hold large pools of investment money derived from retirement funds. They are geared toward long-term investments that will support the desired growth of the fund for an extended time period.

- Retail investors: These are individuals who control a brokerage account and do their own research and make their own decisions on what to buy and sell in the market.

As you can see, there is a great deal of competition for profits in the equities market. Most large institutions spend a lot of money on research that can give them an edge on the future moves of the market. This type of analytical approach depends on accurate information and instant access to trading data, which is possible only with the computational power provided by computer software, most of it written in languages such as C++.

In the next few sections, I provide C++ examples for common problems found in the analysis of equity investments. You will learn about tools and concepts that can be used in a large number of situations in which equity investments are involved.

Moving Average Calculation

Problem

Given a particular equity investment, determine the simple moving average and the exponential moving average for a sequence of closing prices.

Solution

One of the most common strategies to analyze equity instruments such as stocks and ETFs is to use supply/demand methods that consider price and volume as the important variables to observe. Traders who use price/volume-based strategies call this set of methods technical analysis (TA). With TA, traders look at special price points that have been defined by previous price movements, such as support, resistance, trend lines, and moving averages, with the objective of identifying pricing regions with a higher probability of profit.

For example, support and resistance values are typically used to determine price areas that are considered to be of importance for a given instrument. If a stock reaches a certain price when moving up and reverses course, the high price point is considered to be a resistance price. In the future, when the price again reaches the same area, traders will tend to sell around in the same region, creating an even stronger resistance point. Similarly, support prices are formed when traders buy the same stock or ETF in a well-known region.

A similar type of pattern occurs with moving averages. Buyers and sellers tend to look at moving averages to determine if a particular stock is on a low-risk buy or sell point. These psychological price points are self-reinforcing and play an important role in the dynamics of equity trading. Figure 2-1 shows an example of a moving average used in the analysis of common stock for Apple.

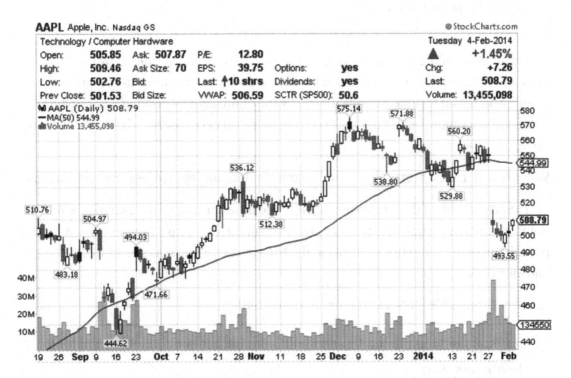

Figure 2-1. *Simple moving average for daily prices of Apple (AAPL), with parameter 50*

The moving average can be calculated using a simple average formula that is repeated for each new period. Given prices p1, p2, ..., pN, the general formula for a particular time period is given by

```
MA = (1/N) (p1 + p2 + ... + pN)
```

You can easily perform this calculation if you maintain and update the sequence of prices as new values are added to the sequence.

To calculate the moving average in C++, we first create a new class that stores a sequence of prices using a STL (standard templates library) vector object. The object is responsible for adding new values to the sequence, using the `addPriceQuote` member function. The implementation of this member function is simple because it relies on the functionality provided by `std::vector` to maintain a sequence of numbers, as well as the storage requirements.

```
void MACalculator::addPriceQuote(double close)
{
    m_prices.push_back(close);
}
```

The number of periods for moving average calculation is determined by the parameter to the constructor of the MACalculator class. For example, to compute a moving average for 20 time periods (normally the equivalent to 4 trading weeks when the period is a single trading day), you can create an object of the MACalculator class in the following way:

```
MACalculator calculator(20); // will compute the moving average for 20 periods.
```

The calculation of the simple moving average is performed by the calculateMA member function of the MACalculator class. The main idea of this function is to iterate through the sequence of prices stored in the MACalculator class, as shown in the following code:

```
std::vector<double> MACalculator::calculateMA()
{
    std::vector<double> ma;
    double sum = 0;
    for (int i=0; i<m_prices.size(); ++i)
    {
        sum += m_prices[i];
        if (i >= m_numPeriods)
        {
            ma.push_back(sum / m_numPeriods);
            sum -= m_prices[i-m_numPeriods];
        }
    }
    return ma;
}
```

To calculate a moving average, it is necessary to have at least the number of observations determined by the number of periods (let N be the number of periods). Therefore, the first N elements of the vector of prices don't generate a corresponding moving average. These initial elements are simply added to the sum local variable, so their values are used later.

For each element after the Nth position, it is possible to calculate the moving average. This is achieved using the sum of the previous N elements and dividing it by the value N. The resulting value is appended to the vector of moving average elements. Finally, it is necessary to update the variable sum, so that the first item of the N-element sequence is dropped from the summation. This happens when the algorithm subtracts the value m_prices[i-m_numPeriods], preparing for the next iteration.

The exponential moving average (EMA) is different from the simple moving average because each new value is multiplied by a factor. This factor is used to give more weight to new values, as compared to older observations. As a result, the EMA is more responsive to changes in the observed values, and it can indicate new trends sooner and with better accuracy. This may be an advantage if you want to quickly spot changes in trend. The following is the code that I used:

```
std::vector<double> MACalculator::calculateEMA()
{
    std::vector<double> ema;
    double multiplier = 2.0 / (m_numPeriods + 1);

    // calculate the MA to determine the first element corresponding
    // to the given number of periods
    std::vector<double> ma = calculateMA();
    ema.push_back(ma.front());

    // for each remaining element, compute the weighted average
    for (int i=m_numPeriods+1; i<m_prices.size(); ++i)
    {
        double val = (1-multiplier) * ema.back() + multiplier * m_prices[i];
        ema.push_back(val);
    }
    return ema;
}
```

The initial part of the calculation is similar to the simple moving average. Values are added using the sum variable, until at least N values have been observed. This is used as the initial value for the EMA. Different implementations of EMA use other ways to initialize the sequence, but the results converge to the same values after a few iterations. You can see a graphical example of EMA in Figure 2-2.

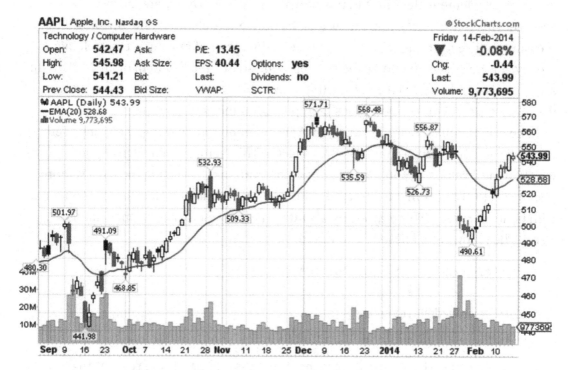

Figure 2-2. *Exponential moving average with a parameter of 20 days*

The main step of the EMA calculation is the addition of new values that are weighted by the multiplier. The default multiplier *r* for EMA computation is given by

$$\rho = \frac{2}{N+1}$$

This multiplier gives greater weight to new values, thus making the EMA more responsive to price changes than the simple moving average.

Complete Code

In Listing 2-1, you can see the complete implementation of the simple moving average as well as the EMA. I also show a sample main function that is responsible for reading a few data points from standard input and calculate the corresponding moving averages.

Listing 2-1. MACalculator

```
//
// MACalculator.h

#ifndef __FinancialSamples__MACalculator__
#define __FinancialSamples__MACalculator__

#include <vector>

class MACalculator {
public:
    MACalculator(int period);
    MACalculator(const MACalculator &);
    MACalculator &operator = (const MACalculator &);
    ~MACalculator();

    void addPriceQuote(double close);
    std::vector<double> calculateMA();
    std::vector<double> calculateEMA();
private:
    // number of periods used in the calculation
    int m_numPeriods;
    std::vector<double> m_prices;
};

#endif /* defined(__FinancialSamples__MACalculator__) */

//
// MACalculator.cpp

#include "MACalculator.h"

#include <iostream>

MACalculator::MACalculator(int numPeriods)
: m_numPeriods(numPeriods)
{

}
```

```cpp
MACalculator::~MACalculator()
{

}

MACalculator::MACalculator(const MACalculator &ma)
: m_numPeriods(ma.m_numPeriods)
{

}

MACalculator &MACalculator::operator = (const MACalculator &ma)
{
    if (this != &ma)
    {
        m_numPeriods = ma.m_numPeriods;
        m_prices = ma.m_prices;
    }
    return *this;
}

std::vector<double> MACalculator::calculateMA()
{
    std::vector<double> ma;
    double sum = 0;
    for (int i=0; i<m_prices.size(); ++i)
    {
        sum += m_prices[i];
        if (i >= m_numPeriods)
        {
            ma.push_back(sum / m_numPeriods);
            sum -= m_prices[i-m_numPeriods];
        }

    }
    return ma;
}
```

```cpp
std::vector<double> MACalculator::calculateEMA()
{
    std::vector<double> ema;
    double sum = 0;
    double multiplier = 2.0 / (m_numPeriods + 1);
    for (int i=0; i<m_prices.size(); ++i)
    {
        sum += m_prices[i];
        if (i == m_numPeriods)
        {
            ema.push_back(sum / m_numPeriods);
            sum -= m_prices[i-m_numPeriods];
        }
        else if (i > m_numPeriods)
        {
            double val = (1-multiplier) * ema.back() + multiplier * m_
            prices[i];
            ema.push_back(val);
        }
    }
    return ema;
}

void MACalculator::addPriceQuote(double close)
{
    m_prices.push_back(close);
}

//
//   main.cpp

#include "MACalculator.h"

#include <iostream>

// the main function receives parameters passed to the program
// and calls the MACalculator class
int main(int argc, const char * argv[])
```

```cpp
{
    if (argc != 2)
    {
        std::cout << "usage: progName <num periods>" << std::endl;
        return 1;
    }

    int num_periods = atoi(argv[1]);

    double price;
    MACalculator calculator(num_periods);
    for (;;) {
        std::cin >> price;
        if (price == -1)
            break;
        calculator.addPriceQuote(price);
    }

    std::vector<double> ma = calculator.calculateMA();

    for (int i=0; i<ma.size(); ++i)
    {
        std::cout << "average value " << i << " = " << ma[i] << std::endl;
    }

    std::vector<double> ema = calculator.calculateEMA();

    for (int i=0; i<ema.size(); ++i)
    {
        std::cout << "exponential average value "
            << i << " = " << ema[i] << std::endl;
    }

    return 0;
}
```

Running the Code

You can compile this code using the gcc compiler (as well as any other standards-compliant compiler such as Visual Studio or C++ builder). For example, the following command line can be used from the UNIX shell:

```
gcc -o macalc main.cpp macalculator.cpp
```

The following is a display of a sample execution of the program:

```
$ ./macalc 5
10
11
22
12
13
23
12
32
12
3
2
22
32
-1
average value 0 = 18.2
average value 1 = 18.6
average value 2 = 22.8
average value 3 = 20.8
average value 4 = 19
average value 5 = 16.8
average value 6 = 16.6
average value 7 = 20.6
exponential average value 0 = 18.2
exponential average value 1 = 16.1333
exponential average value 2 = 21.4222
exponential average value 3 = 18.2815
```

47

```
exponential average value 4 = 13.1877
exponential average value 5 = 9.45844
exponential average value 6 = 13.639
exponential average value 7 = 19.7593
Program ended with exit code: 0
```

In the first line, I entered the command that calls the moving average program (which here is simply called `macalc`). The single argument in the command line means that I want to calculate the moving average for five data points. Then, I entered a sequence of numbers that represents the observed prices for a certain investment vehicle. Finally, I entered the value -1, which indicates the end of the input. The next few lines then give a list of values that define the simple moving average and the EMA.

Calculating Volatility

Problem

Calculate the volatility of a particular equity instrument, given a sequence of prices for the last few days.

Solution

One of the important characteristics of stocks and other equity instruments is that they change in price very frequently. For highly liquid stocks and ETFs, prices will change during the whole trading day, as new buyers and sellers exchange shares. The result is a high degree of volatility, as compared to other investment instruments.

Volatility is also an important concept when comparing investment options. For example, an Internet stock will vary in price much more widely than a traditional food producer. Their volatility profiles will be completely different. Higher volatility may be an advantage or a disadvantage, depending on your investment objectives.

The important thing to consider about volatility is that it is not just a one-dimensional concept. Different investment strategies require different ways of viewing price variations. For example, if you are making investment decisions based on the expected volatility for the next few days (due to a news event or earnings release), then the previous week's volatility may not be so important.

In this section, I present three ways to measure volatility given a sequence of prices. The first strategy is computing the range of values observed during that period. This is probably the simplest way to view volatility: calculate the highest and lowest observed values and return its difference. It is also a common indicator used by many investors. Most newspapers print a list of 1-year high and low prices, so you can quickly see the simple range for the previous year. The following is the implementation using a vector of prices:

```
double VolatilityCalculator::rangeVolatility()
{
    if (m_prices.size() < 1)
    {
        return 0;
    }

    double min = m_prices[0];
    double max = min;
    for (int i=1; i<m_prices.size(); ++i)
    {
        if (m_prices[i] < min)
        {
            min = m_prices[i];
        }
        if (m_prices[i] > max)
        {
            max = m_prices[i];
        }
    }
    return max - min;
}
```

The second strategy is calculating the average range for a given time period. For example, many investment strategies use the idea of looking at the past few days and taking an average of the observed ranges. The result is then charted as an indicator of the rate of change for a particular stock, for example. Simply calculating the average of the previously observed daily ranges can be used to return this value. Here is our code.

```
double VolatilityCalculator::avgDailyRange()
{
    unsigned long n = m_prices.size();
    if (n < 2)
    {
        return 0;
    }

    double previous = m_prices[0];
    double sum = 0;
    for (int i=1; i<m_prices.size(); ++i)
    {
        double range = abs(m_prices[i] - previous);
        sum += range;
    }
    return sum / n - 1;
}
```

Finally, a more sophisticated way to gauge the variation of values for an equity instrument is to use the statistical definition of standard deviation. The standard deviation is useful as a way to derive volatility from the expected value (also known as mean) of a set of prices. A well-known formula is used to calculate the standard deviation, which is given by

$$\frac{1}{N-1}\sum_{i=1}^{N}(x_i - \mu)^2$$

In this equation, N is the number of data points (prices) and m is the average of these values. The standard deviation can be calculated in C++ with the following code:

```
double VolatilityCalculator::stdDev()
{
    const double m = mean();
    double sum = 0;
    for (int i=0; i<m_prices.size(); ++i)
    {
        double val = m_prices[i] - m;
```

```
        sum += val * val;
    }
    return sqrt(sum / (m_prices.size()-1));
}
```

Complete Code

Listing 2-2 provides the complete code for the strategies just described. I introduce a new C++ class named VolatilityCalculator, which encapsulates the concept of computing the price volatility. We have these three strategies coded in the rangeVolatility, avgDailyRange, and stdDev member functions. You can use this class as a starting point and later add other methods for volatility calculation as additional member functions.

Listing 2-2.   VolatilityCalculator.h

```
//
// VolatilityCalculator.h

#ifndef __FinancialSamples__VolatilityCalculator__
#define __FinancialSamples__VolatilityCalculator__

#include <vector>

class VolatilityCalculator
{
public:
    VolatilityCalculator();
    ~VolatilityCalculator();
    VolatilityCalculator(const VolatilityCalculator &);
    VolatilityCalculator &operator=(const VolatilityCalculator &);

    void addPrice(double price);
    double rangeVolatility();
    double stdDev();
    double mean();
    double avgDailyRange();
```

```cpp
private:
    std::vector<double> m_prices;
};

#endif /* defined(__FinancialSamples__VolatilityCalculator__) */

//
//  VolatilityCalculator.cpp

#include "VolatilityCalculator.h"

#include <iostream>
#include <cmath>

VolatilityCalculator::VolatilityCalculator()
{

}

VolatilityCalculator::~VolatilityCalculator()
{

}

VolatilityCalculator::VolatilityCalculator(const VolatilityCalculator &v)
: m_prices(v.m_prices)
{

}

VolatilityCalculator &VolatilityCalculator::operator =(const
VolatilityCalculator &v)
{
    if (&v != this)
    {
        m_prices = v.m_prices;
    }
    return *this;
}
```

```cpp
void VolatilityCalculator::addPrice(double price)
{
    m_prices.push_back(price);
}

double VolatilityCalculator::rangeVolatility()
{
    if (m_prices.size() < 1)
    {
        return 0;
    }

    double min = m_prices[0];
    double max = min;
    for (int i=1; i<m_prices.size(); ++i)
    {
        if (m_prices[i] < min)
        {
            min = m_prices[i];
        }
        if (m_prices[i] > max)
        {
            max = m_prices[i];
        }
    }
    return max - min;
}

double VolatilityCalculator::avgDailyRange()
{
    unsigned long n = m_prices.size();
    if (n < 2)
    {
        return 0;
    }
```

```cpp
    double previous = m_prices[0];
    double sum = 0;
    for (int i=1; i<m_prices.size(); ++i)
    {
        double range = abs(m_prices[i] - previous);
        sum += range;
    }
    return sum / n - 1;
}

double VolatilityCalculator::mean()
{
    double sum = 0;
    for (int i=0; i<m_prices.size(); ++i)
    {
        sum += m_prices[i];
    }
    return sum/m_prices.size();
}

double VolatilityCalculator::stdDev()
{
    double m = mean();
    double sum = 0;
    for (int i=0; i<m_prices.size(); ++i)
    {
        double val = m_prices[i] - m;
        sum += val * val;
    }
    return sqrt(sum / (m_prices.size()-1));
}
//
//  main.cpp

#include "VolatilityCalculator.h"

#include <iostream>
```

```cpp
// the main function receives parameters passed to the program
int main(int argc, const char * argv[])
{
    double price;

    VolatilityCalculator vc;
    for (;;)
    {
        std::cin >> price;
        if (price == -1)
        {
            break;
        }
        vc.addPrice(price);
    }

    std::cout << "range volatility is " <<  vc.rangeVolatility()  <<
    std::endl;
    std::cout << "average daily range is " <<  vc.avgDailyRange()  <<
    std::endl;
    std::cout << "standard deviation is " <<  vc.stdDev()  << std::endl;
    return 0;
}
```

Running the Code

Here is an example of the volatility class being used. You can compile the code presented in Listing 2-2, assuming that the binary is called volatility. Then, you can use the program by entering price values that will be later used to compute the volatility employing the three methods described. The end of the input sequence is determined by a single -1 value entered as the last input value.

```
$ ./volatility
3
3.5
5
4.48
```

```
5.2
6
6.1
5.5
5.2
5.7
-1
range volatility is 3.1
average daily range is 0.7
standard deviation is 1.02957
```

Computing Instrument Correlation

Problem

Given a sequence of closing prices for the last N periods, calculate the correlation between two equity instruments.

Solution

One of the main problems that money managers need to solve is how to diversify a portfolio. The problem of diversification occurs because, when investing in the market, it is not desirable to have all your assets in the same type of investment. Correlated investments tend to go down at the same time, making it harder to avoid losses in a portfolio.

For example, consider two companies operating in a similar business. The classic example is beverage companies such as Coca-Cola and Pepsi. They tend to rise and fall at the same time due to the similarity of their business. Therefore, we say that they are highly correlated. Correlation is a mathematical concept that was developed for the analysis of statistical events. It turns out to be an important concept in the equities market, since probability plays such a big role in the evaluation and modeling of equity-based investments.

To make the code for this example more extensible, we divide the solution into two classes. The first class, called TimeSeries, represents the often-used concept of a set of numbers that apply to a certain quantity over a given period of time. This concept is commonly referred to as a time series. The TimeSeries class is responsible for calculating values that are specific to a single time series, such as the average, or the standard deviation.

The second class used is CorrelationCalculator, which is responsible for collecting data for the desired time series and computing the correlation using the formula

$$\frac{1}{N-1}\frac{\Sigma(x_i-\bar{x})(y_i-\bar{y})}{\sigma_x\sigma_y}$$

In this equation, N is the number of observations, x_i is the i-th observation of the first time-series, y_i is the i-th observation of the second time-series, $\bar{x}$ and $\bar{y}$ are the mean (average) of the two sequences of prices, s_x is the standard deviation of the x values, and s_y is the standard deviation of the y values.

The mean value and the standard deviation are calculated in the TimeSeries class. These values are then used in the CorrelationCalculator to determine the correlation between the values observed for both sequences.

Complete Code

The computation discussed in the previous section is implemented in the class TimeSeries. Listing 2-3 includes the complete class. You can also see how to use techniques to calculate correlation, as displayed in the class CorrelationCalculator.

Listing 2-3.   TimeSeries.h

```
//
//  TimeSeries.h

#ifndef __FinancialSamples__TimeSeries__
#define __FinancialSamples__TimeSeries__

#include <vector>

class TimeSeries
{
public:
    TimeSeries();
```

```cpp
    TimeSeries(const TimeSeries &);
    TimeSeries &operator=(const TimeSeries &);
    ~TimeSeries();

    void addValue(double val);
    double stdDev();
    double mean();
    size_t size();
    double elem(int i);
private:
    std::vector<double> m_values;
};

#endif /* defined(__FinancialSamples__TimeSeries__) */

//
//   TimeSeries.cpp

#include "TimeSeries.h"
#include <cmath>
#include <iostream>

TimeSeries::TimeSeries()
: m_values()
{

}

TimeSeries::~TimeSeries()
{

}

TimeSeries::TimeSeries(const TimeSeries &ts)
: m_values(ts.m_values)
{
}

TimeSeries &TimeSeries::operator =(const TimeSeries &ts)
{
```

```
    if (this != &ts)
    {
        m_values = ts.m_values;
    }
    return *this;
}

void TimeSeries::addValue(double val)
{
    m_values.push_back(val);
}

double TimeSeries::mean()
{
    double sum = 0;
    for (int i=0; i<m_values.size(); ++i)
    {
        sum += m_values[i];
    }
    return sum/m_values.size();
}

double TimeSeries::stdDev()
{
    double m = mean();
    double sum = 0;
    for (int i=0; i<m_values.size(); ++i)
    {
        double val = m_values[i] - m;
        sum += val * val;
    }

    return sqrt(sum / (m_values.size()-1));
}

size_t TimeSeries::size()
{
    return m_values.size();
}
```

```cpp
double TimeSeries::elem(int pos)
{
    return m_values[pos];
}

//
// CorrelationCalculator.h

#ifndef __FinancialSamples__CorrelationCalculator__
#define __FinancialSamples__CorrelationCalculator__

class TimeSeries;

class CorrelationCalculator
{
public:
    CorrelationCalculator(TimeSeries &a, TimeSeries &b);
    ~CorrelationCalculator();
    CorrelationCalculator(const CorrelationCalculator &);
    CorrelationCalculator &operator =(const CorrelationCalculator &);

    double correlation();
private:
    TimeSeries &m_tsA;
    TimeSeries &m_tsB;
};

#endif /* defined(__FinancialSamples__CorrelationCalculator__) */

//
// CorrelationCalculator.cpp

#include "CorrelationCalculator.h"

#include "TimeSeries.h"
#include <iostream>

CorrelationCalculator::CorrelationCalculator(TimeSeries &a, TimeSeries &b)
: m_tsA(a),
  m_tsB(b)
```

```cpp
{
}

CorrelationCalculator::~CorrelationCalculator()
{

}

CorrelationCalculator::CorrelationCalculator(const CorrelationCalculator &c)
: m_tsA(c.m_tsA),
  m_tsB(c.m_tsB)
{

}

CorrelationCalculator &CorrelationCalculator::operator=(const
CorrelationCalculator &c)
{
    if (this != &c)
    {
        m_tsA = c.m_tsA;
        m_tsB = c.m_tsB;
    }
    return *this;
}

double CorrelationCalculator::correlation()
{
    double sum = 0;
    double meanA = m_tsA.mean();
    double meanB = m_tsB.mean();

    if (m_tsA.size() != m_tsB.size()) {
        std::cout << "error: number of observations is different"
        << std::endl;
        return -1;
    }
```

```cpp
    for (int i=0; i<m_tsA.size(); ++i)
    {
        auto val = (m_tsA.elem(i) - meanA) * (m_tsB.elem(i) - meanB);
        sum += val;
    }
    double stDevA = m_tsA.stdDev();
    double stDevB = m_tsB.stdDev();
    sum /= (stDevA * stDevB);
    return sum / (m_tsB.size() - 1);
}

//
//  main.cpp

#include "CorrelationCalculator.h"
#include "TimeSeries.h"

#include <iostream>

// the main function receives parameters passed to the program
int main(int argc, const char * argv[])
{
    double price;

    TimeSeries tsa;
    TimeSeries tsb;
    for (;;) {
        std::cin >> price;
        if (price == -1)
        {
            break;
        }
        tsa.addValue(price);
        std::cin >> price;
        tsb.addValue(price);
    }

    CorrelationCalculator cCalc(tsa, tsb);
```

```
auto correlation = cCalc.correlation();

std::cout << "correlation is " <<  correlation  << std::endl;
return 0;
}
```

Running the Code

After compiling the provided code, you can run the resulting program by calling the executable without any parameters. The program works by reading the data from standard input, which you can do manually or by redirecting a file to the program using the shell. Each line of the input contains prices for the two equity instruments we want to compare. The last line is marked using the special value -1, which indicates the end of the input stream.

The following is a sample execution:

```
$ ./correlation
1.2 3.4
2 3.3
2.5 3
4 5.5
3 1.2
6 2.4
5.5 3.2
6.3 3.1
7.1 2.9
5.4 3.2
-1
correlation is -0.050601
```

The second example shows the result for stocks that display inverse correlation: when the price of the first instrument increases, the price of the second one decreases.

```
$ ./correlation
1 10
2 9
3 8
```

```
4 7
5 6
6 5
7 4
-1
 avg is 4
 avg is 7
 avg is 4
 avg is 7
correlation is -1
```

Calculating Fundamental Indicators

Problem

Compute a set of fundamental indicators for a particular stock holding.

Solution

In the last few sections, we have seen methods for analyzing price changes in equity instruments. These techniques are generally labeled as technical indicators, since they allow for the TA of past price and volume data. Another way to analyze stocks is to consider more fundamental information that is not contained in the sequence of observed prices. Such fundamental information includes company earnings, intellectual property, physical assets, and debt.

Fundamental indicators are one of the most common ways of analyzing the quality of a stock. The disclosure of fundamental information is required from public companies and released every quarter for most publicly traded stocks. It includes financial data that is considered by the Securities and Exchange Commission to be of value for investors and is used to tell how well a company is performing compared to its peers in the marketplace. For example, *earnings per share* are a fundamental indicator that tells how much profit is being generated per period (usually a quarter or a year) for each share of the stock. This information is then used to make decisions about buying, selling, or holding a particular investment vehicle.

In this example, I present a class that can be used to model stocks and allows one to calculate and display a set of fundamental indicators associated with the stock. The code is encapsulated in the class FundamentalsCalculator. The idea is to have a central location where you can calculate and store all the fundamental indicators associated with a stock.

Here is a list of the items that you can retrieve using the FundamentalsCalculator class and how they are defined:

> Price-earnings ratio (P/E): This is calculated as the price of the total stock of the company divided by the earnings as published in the last-quarter earnings release. This ratio can be interpreted as a measure of the cost of the company stock as compared to other companies with similar earnings.

> Book value: The book value corresponds to the amount of assets currently on the company balance sheet. This is in essence an accounting measure of the value of the company, without considering market factors such as future earnings, for example.

> Price-to-book ratio (P/B): This ratio is determined by dividing the stock price by the assets minus liabilities. The following accounting formula can be used:

$$\frac{StockPrice}{Assets - Liablities \ and \ IntangibleAssets}$$

> Notice that only tangible assets, the ones that can be eventually sold, are considered in this equation.

> Price-earnings to growth (PEG): This indicator can be used to compare companies with similar P/E but different growth rates. The formula to calculate this value is simply

$$\frac{\frac{P}{E}}{EPS \ annual \ growth}$$

> Earnings before interest, taxes, depreciation, and amortization (EBITDA): This is a measure that can be used to determine how a company is making a profit, and it is based on accounting information provided by the company in every earnings release.

The value simply represents how much profit the company made before items such as taxes and related expenses were paid.

Return on equity (ROE): This ratio is used to determine the percentage of net income generated based on shareholders' equity. Investors are usually interested in companies able to generate higher income on the same amount of equity. The value is simply calculated as

$$\frac{NetIncome}{ShareholdersEquity}$$

Forward P/E: This number is similar to the P/E ratio, but instead of being calculated based on existing revenue data, it is a prediction for the next quarter made by analysts. When compared to P/E, this number can be used to determine if analysts expect the revenue to increase, decrease, or stay at the same levels.

Complete Code

Most of the indicators explained in the previous list are easy to calculate, but they are very important when making decisions on which stocks to buy or sell. The class presented in Listing 2-2 offers a good place to store the associated data needed for these indicators, along with the simple calculations needed to produce the desired values with the minimum amount of input.

Listing 2-4 is the complete listing of the FundamentalsCalc class and its associated test code.

Listing 2-4. FundamentalsCalc.h

```
//
// FundamentalsCalc.h

#ifndef __FinancialSamples__FundamentalsCalc__
#define __FinancialSamples__FundamentalsCalc__

#include <string>

class FundamentalsCalculator {
```

```cpp
public:
    FundamentalsCalculator(const std::string &ticker, double price, double
    dividend);
    ~FundamentalsCalculator();
    FundamentalsCalculator(const FundamentalsCalculator &);
    FundamentalsCalculator &operator=(const FundamentalsCalculator&);

    void setNumOfShares(int n);
    void setEarnings(double val);
    void setExpectedEarnings(double val);
    void setBookValue(double val);
    void setAssets(double val);
    void setLiabilitiesAndIntangibles(double val);
    void setEpsGrowth(double val);
    void setNetIncome(double val);
    void setShareHoldersEquity(double val);

    double PE();
    double forwardPE();
    double bookValue();
    double priceToBookRatio();
    double priceEarningsToGrowth();
    double returnOnEquity();
    double getDividend();

private:

    std::string m_ticker;
    double m_price;
    double m_dividend;
    double m_earningsEstimate;
    int m_numShares;
    double m_earnings;
    double m_bookValue;
    double m_assets;
    double m_liabilitiesAndIntangibles;
    double m_epsGrowth;
```

```cpp
    double m_netIncome;
    double m_shareholdersEquity;
};

#endif /* defined(__FinancialSamples__FundamentalsCalc__) */

//
//  FundamentalsCalc.cpp

#include "FundamentalsCalc.h"

#include <iostream>

FundamentalsCalculator::FundamentalsCalculator(const std::string &ticker,
                                               double price, double
                                               dividend) :
m_ticker(ticker),
m_price(price),
m_dividend(dividend),
m_earningsEstimate(0),
m_numShares(0),
m_bookValue(0),
m_assets(0),
m_liabilitiesAndIntangibles(0),
m_epsGrowth(0),
m_netIncome(0),
m_shareholdersEquity(0)
{

}

FundamentalsCalculator::FundamentalsCalculator(const FundamentalsCalculator &v) :
m_ticker(v.m_ticker),
m_price(v.m_price),
m_dividend(v.m_dividend),
m_earningsEstimate(v.m_earningsEstimate),
m_numShares(v.m_numShares),
m_bookValue(v.m_bookValue),
m_assets(v.m_assets),
```

```
m_liabilitiesAndIntangibles(v.m_liabilitiesAndIntangibles),
m_epsGrowth(v.m_epsGrowth),
m_netIncome(v.m_netIncome),
m_shareholdersEquity(v.m_shareholdersEquity)
{

}

FundamentalsCalculator::~FundamentalsCalculator()
{

}

FundamentalsCalculator &FundamentalsCalculator::operator=(const
FundamentalsCalculator &v)
{
    if (this != &v)
    {
        m_ticker = v.m_ticker;
        m_price = v.m_price;
        m_dividend = v.m_dividend;
        m_earningsEstimate = v.m_earningsEstimate;
        m_numShares = v.m_numShares;
        m_bookValue = v.m_bookValue;
        m_assets = v.m_assets;
        m_liabilitiesAndIntangibles = v.m_liabilitiesAndIntangibles;
        m_epsGrowth = v.m_epsGrowth;
        m_netIncome = v.m_netIncome;
        m_shareholdersEquity = v.m_shareholdersEquity;
    }
    return *this;
}

double FundamentalsCalculator::PE()
{
    return (m_price * m_numShares)/ m_earnings;
}
```

```
double FundamentalsCalculator::forwardPE()
{
    return (m_price * m_numShares)/ m_earningsEstimate;
}

double FundamentalsCalculator::returnOnEquity()
{
    return m_netIncome / m_shareholdersEquity;
}

double FundamentalsCalculator::getDividend()
{
    return m_dividend;
}

double FundamentalsCalculator::bookValue()
{
    return m_bookValue;
}

double FundamentalsCalculator::priceToBookRatio()
{
    return (m_price * m_numShares) / (m_assets -
    m_liabilitiesAndIntangibles);
}

double FundamentalsCalculator::priceEarningsToGrowth()
{
    return PE()/ m_epsGrowth;
}

void FundamentalsCalculator::setNumOfShares(int n)
{
    m_numShares = n;
}

void FundamentalsCalculator::setEarnings(double val)
{
    m_earnings = val;
```

```
}

void FundamentalsCalculator::setExpectedEarnings(double val)
{
    m_earningsEstimate = val;
}

void FundamentalsCalculator::setBookValue(double val)
{
    m_bookValue = val;
}

void FundamentalsCalculator::setEpsGrowth(double val)
{
    m_epsGrowth = val;
}

void FundamentalsCalculator::setNetIncome(double val)
{
    m_netIncome = val;
}

void FundamentalsCalculator::setShareHoldersEquity(double val)
{
    m_shareholdersEquity = val;
}

void FundamentalsCalculator::setLiabilitiesAndIntangibles(double val)
{
    m_liabilitiesAndIntangibles = val;
}

void FundamentalsCalculator::setAssets(double val)
{
    m_assets = val;
}
//
//  main.cpp
```

```cpp
#include "FundamentalsCalc.h"

#include <iostream>

// the main function receives parameters passed to the program
// and uses class FundamentalsCalculator
int main(int argc, const char * argv[])
{
    FundamentalsCalculator fc("AAPL", 543.99, 12.20);

    // values are in millions
    fc.setAssets(243139);
    fc.setBookValue(165234);
    fc.setEarnings(35885);
    fc.setEpsGrowth(0.22);
    fc.setExpectedEarnings(39435);
    fc.setLiabilitiesAndIntangibles(124642);
    fc.setNetIncome(37235);
    fc.setNumOfShares(891990);
    fc.setShareHoldersEquity(123549);

    std::cout << "P/E: " <<  fc.PE()/1000  << std::endl;  // prices in thousands
    std::cout << "forward P/E: " <<  fc.forwardPE()/1000  << std::endl;
    std::cout << "book value: " <<  fc.bookValue()  << std::endl;
    std::cout << "price to book: " <<  fc.priceToBookRatio()  << std::endl;
    std::cout << "price earnings to growth: " <<  fc.
priceEarningsToGrowth()  << std::endl;
    std::cout << "return on equity: " <<  fc.returnOnEquity()  << std::endl;
    std::cout << "dividend: " <<  fc.getDividend()  << std::endl;
    return 0;
}
```

Running the Code

You can compile the code displayed in Listing 2-4 along with the respective test contained in the main function. The result would be displayed as follows:

```
$ ./fundamentalind
P/E: 13.5219
forward P/E: 12.3046
book value: 165234
price to book: 4094.9
price earnings to growth: 61463.2
return on equity: 0.301378
dividend: 12.2
```

Conclusion

In this chapter, I provided an overview of the problems and opportunities in the equities market. As you have seen, being a major part of the financial system, equity trading is an area in which computational problems exist in all phases of analysis and trade execution.

The chapter starts with a short introduction to the equities market, describing the main players and the financial instruments used in the trading process. In the first section, you learned how to calculate moving averages using C++. Moving averages are widely used to uncover trends in stock prices. In the same section, I discussed how to calculate the EMA, in which the most recent prices receive a larger weight. The EMA is more responsive to recent changes in price, which may be a better way to make buy or sell decisions in some algorithms.

Next, I presented some code to calculate the volatility of an equity instrument. The notion of volatility is important when making decisions about which instruments to hold in a portfolio. The methods for calculating volatility include using the simple observed range as well as the probabilistic measure of volatility, also called standard deviation.

In this chapter, you have also learned how to calculate the correlation between two stocks, indicating if there is positive, negative, or no correlation based on their observed prices. Finally, this chapter introduces techniques for modeling and calculating fundamental data about a stock holding. Such a C++ class is easy to create, but it is also very useful when fundamental data is required during the analysis of a particular stock. You can modify this example to add new fundamental indicators as needed and therefore reuse existing code in other areas of your financial applications.

In the next chapter, you will learn more about C++ features that are frequently used in the creation of financial software. You will see a number of techniques that are readily available to developers in the financial industry. Such C++ features are able to improve the performance, robustness, and flexibility of most code that is created for the analysis of investments.

CHAPTER 3

C++ Programming Techniques in Finance

The C++ language was designed as an extension of C, which means that most programs written in C are also valid C++ programs. However, experienced programmers typically make use of a set of high-level features made available exclusively in C++ as a way to control program complexity, including features that were introduced in the C++20 standard. This is an especially important consideration for financial software development, where we want to create fast and expressive applications.

In this chapter, we explore a few fundamental techniques that financial programmers have used over the years to write better C++ code with less effort. These techniques have been selected among the many features provided by C++ as the most effective in improving the quality and expressiveness of code. Such features include the following:

- Templates: A feature that allows the creation of generic software, with classes and functions that can be applied over a set of possibly unrelated types that satisfy the set of requirements for a desired operation.

- Shared pointers: A programming technique that reduces the need for direct manipulation of pointers. With shared pointers, you can avoid a big source of mistakes inherent to the way C++ programs manage memory and other resources.

- Operator overloading: With overloading, you can apply standard operators already available in the language to your own classes and structures.

- C++20 features: The latest iteration of the C++ standard has introduced many new features that help control the complexity of programs. These features, which can be easily used for the creation of financial software, include shared pointers and automatic type detection.

© Carlos Oliveira 2021
C. Oliveira, *Practical C++20 Financial Programming*, https://doi.org/10.1007/978-1-4842-6834-6_3

In the next sections, you will see a few selected programming examples that explore some of these C++ features in the context of financial applications.

Calculating Interest Rates for Investment Instruments

Interest rates are a fundamental concept for fixed income investors. Design a C++ solution to return the annual interest rate, given a generic instrument class that provides methods such as getMonthlyPayment and getPrincipal.

Solution

The foregoing problem is frequently used in the design of interest rate calculation engines. You can create a solution using a number of strategies such as class hierarchies, but for performance and design considerations, the use of templates is the most indicated method of combining interest rate data from unrelated classes that represent investment instruments.

A *template* is a mechanism, along with a special syntax, used to create code that works with different underlying data types. Using templates, one can create functions, member functions, and classes that are able to support different types using the same code. The code generated using template-based programming techniques is said to be generic, since it can be used with different types (either fundamental types such as int or double or user-defined classes and structures). Generic functions and types are instantiated using the name of the target type(s) between angle brackets, which indicates a particular version of the desired function or type. In the most recent standard revisions, a generic function can also be defined using automatic argument deduction.

The creation and use of generic code are possible because when the compiler finds a template, it does not generate code immediately. Instead, code is generated only at the point where the template object or function is instantiated. When that happens, the compiler detects the types involved in the expression, and the template is instantiated. Only then the traditional compilation steps such as syntactic analysis and code generation are performed on the instantiated code, and any resulting errors will be detected and reported to the programmer.

To solve the interest rate calculation problem, you can use templates to implement an interest rate engine class, called `IntRateEngine`. This class is defined in such a way that you can apply it to any class implementing the methods `getMonthlyPayment` and `getPrincipal`. I have included two sample classes that implement these methods, the classes `BondInstrument` and `MortgageInstrument`. However, the big advantage of using templates is that you don't need to derive such classes from a particular base class, for example. You can use a class supplying these same methods, and the compiler will do the hard work of combining these classes. This means that there is no coupling between investment instruments and the interest rate calculation engine. In fact, if you look at the files for `IntRateEngine`, you will not find any reference to the investment instrument classes.

Here is a quick look at the relevant parts of the `BondInstrument` class.

```
class BondInstrument {
public:
    double getMonthlyPayment();
    double getPrincipal();

    // other methods here...
};
```

With this class, one could instantiate the template that calculates the annual interest rate. The template class that performs the calculation is defined in the following way:

```
template <class T>
class IntRateEngine {
public:
    void setInstrument(T &inv);
    double getAnnualIntRate();
    // other methods here ...
private:
    T m_instrument;
};
```

Notice that the type of instrument is left unspecified as a type argument T. This is the parameterization that allows different classes to be used with the same template. Similarly, you can see the implementation of the `getAnnualIntRate` method.

```
template <class T>
double IntRateEngine<T>::getAnnualIntRate()
{
    double payment = m_instrument.getMonthlyPayment();
    double principal = m_instrument.getPrincipal();
    return (12 *payment) / principal;
}
```

Notice that the method only requires the parameter T to be offered the getMonthlyPayment and getPrincipal methods. Any type that supports these two methods can be used by IntRateEngine without problems.

Complete Code

The algorithm described previously has been implemented in the classes BondInstrument, MortgageInstrument, and IntRateEngine, as displayed in Listing 3-1.

Listing 3-1. InvestmentInstrument.h

```
//
//  InvestmentInstrument.h

#ifndef __FinancialSamples__InvestmentInstrument__
#define __FinancialSamples__InvestmentInstrument__

#include <iostream>

class BondInstrument {
public:

    BondInstrument(double principal, double monthlyPayment);
    ~BondInstrument();
    BondInstrument(const BondInstrument &a);
    BondInstrument &operator =(const BondInstrument &a);

    double getMonthlyPayment();
    double getPrincipal();
```

```cpp
    // other methods here...
private:
    double
        m_monthlyPay,
        m_principal;
};

class MortgageInstrument {
public:

    MortgageInstrument(double monthlyPay, double propertyValue, double
    downpayment);
    ~MortgageInstrument();
    MortgageInstrument(const MortgageInstrument &a);
    MortgageInstrument &operator =(const MortgageInstrument &a);

    double getMonthlyPayment();
    double getPrincipal();

    // other methods here...
private:
    double
        m_monthlyPay,
        m_propertyValue,
        m_downPayment;
};

#endif /* defined(__FinancialSamples__InvestmentInstrument__) */

//
//  InvestmentInstrument.cpp

#include "InvestmentInstrument.h"

BondInstrument::BondInstrument(double principal, double monthlyPayment)
: m_principal(principal),
m_monthlyPay(monthlyPayment)
{

}
```

```cpp
BondInstrument::~BondInstrument()
{

}

BondInstrument::BondInstrument(const BondInstrument &a)
: m_monthlyPay(a.m_monthlyPay),
m_principal(a.m_principal)
{

}

BondInstrument &BondInstrument::operator =(const BondInstrument &a)
{
    if (this != &a)
    {
        m_principal = a.m_principal;
        m_monthlyPay = a.m_monthlyPay;
    }
    return *this;
}

double BondInstrument::getMonthlyPayment()
{
    return m_monthlyPay;
}

double BondInstrument::getPrincipal()
{
    return m_principal;
}

/////////////

MortgageInstrument::MortgageInstrument(double monthlyPay, double
propertyValue, double downpayment)
: m_monthlyPay(monthlyPay),
m_propertyValue(propertyValue),
```

```
m_downPayment(downpayment)
{

}

MortgageInstrument::~MortgageInstrument()
{

}

MortgageInstrument::MortgageInstrument(const MortgageInstrument &a)
: m_downPayment(a.m_downPayment),
m_propertyValue(a.m_propertyValue),
m_monthlyPay(a.m_monthlyPay)
{

}

MortgageInstrument &MortgageInstrument::operator =(const MortgageInstrument &a)
{
    if (this != &a)
    {
        m_downPayment = a.m_downPayment;
        m_propertyValue = a.m_propertyValue;
        m_monthlyPay = a.m_monthlyPay;
    }
    return *this;
}

double MortgageInstrument::getMonthlyPayment()
{
    return m_monthlyPay;
}

double MortgageInstrument::getPrincipal()
{
    return m_propertyValue - m_downPayment;
}
```

```
//
//  IntRateEngine.h

#ifndef __FinancialSamples__IntRateEngine__
#define __FinancialSamples__IntRateEngine__

#include <vector>

template <class T>
class IntRateEngine {
public:

    ~IntRateEngine();
    IntRateEngine(const IntRateEngine<T> &a);
    IntRateEngine<T> &operator =(const IntRateEngine<T> &a);

    void setInstrument(T &inv);
    double getAnnualIntRate();
private:
    T m_instrument;
};

template <class T>
IntRateEngine<T>::~IntRateEngine()
{

}

template <class T>
IntRateEngine<T>::IntRateEngine(const IntRateEngine<T> &a)
: m_instrument(a.m_instrument)
{

}

template <class T>
IntRateEngine<T> &IntRateEngine<T>::operator =(const IntRateEngine<T> &a)
{
    if (this != &a)
    {
```

```
        m_instrument = a.m_instrument;
    }
    return *this;
}

template <class T>
void IntRateEngine<T>::setInstrument(T &inv)
{
    m_instrument = inv;
}

template <class T>
double IntRateEngine<T>::getAnnualIntRate()
{
    double payment = m_instrument.getMonthlyPayment();
    double principal = m_instrument.getPrincipal();
    return (12 *payment) / principal;
}

#endif /* defined(__FinancialSamples__IntRateEngine__) */

//
// main.cpp

#include "InvestmentInstrument.h"
#include "IntRateEngine.h"

#include <iostream>

int main()
{
    IntRateEngine<BondInstrument> engineA;
    IntRateEngine<MortgageInstrument> engineB;

    BondInstrument bond(40000, 250);
    MortgageInstrument mortgage(250, 50000, 5000);
    engineA.setInstrument(bond);
    engineB.setInstrument(mortgage);
```

```
std::cout << " bond annual int rate: " << engineA.getAnnualIntRate()*
100 << "%" << std::endl;
std::cout << " mortgage annual int rate: " << engineB.getAnnualIntRate()*
100 << "%" << std::endl;

return 0;
}
```

Running the Code

You can compile the code presented in Listing 3-1 with any standards-compliant compiler. For example, with gcc, you can use the following command line:

```
> gcc -o intRate main.cpp InvestmentInstrument.cpp IntRateEngine.cpp
```

You can use the resulting test program without any parameters.

```
> ./intRate
 bond annual int rate: 7.5%
 mortgage annual int rate: 6.66667%
```

Creating Financial Statement Objects

Create a class that computes a financial statement and returns it to the calling client. Do this while avoiding potential memory leaks of the returned data.

Solution

To solve this problem, you will create a simple financial statement class and a function that returns a pointer to the financial statement. The main issue that you may want to avoid in the solution is the possibility of memory leaks occurring when the financial statement is returned. For this purpose, you will learn about how to use smart pointers to manage memory. Before we explain that, however, you need to understand the concept of smart pointers.

Smart Pointers

The C language popularized the concept of pointers, and it innovated when it provided a simple notation for direct memory access on a high-level language. Pointers allow the manipulation of memory addresses in a way that complies with the underlying data type of the data. For example, it is possible to use pointers to refer to successive addresses using increment and decrement operators. For instance, you may have

```
int numbers[200];
// initialize numbers here ...
//
int *parray = numbers;
for (int i=0; i<200; ++i) {
        std::cout << "value is " << *parray << "\n";
        parray++;
}
```

In the previous example, parray is a pointer to an integer. It can be used to refer to the location of any integer value in memory. In particular, you can use it to hold the address of the first element of the numbers array. The code in the for loop is then used to print the current value pointed by parray and to update its address using the increment operator, which moves the pointer to the next integer location. Notice that this is possible even though the number of bytes per integer is greater than one. In fact, the increment operator is aware of the pointer type used and will change the address to point to the exact location of the next value for the declared type.

While the notation for pointers is very powerful, and fully supported by C++, programmers should avoid the use of pointers in C++ code whenever possible. Pointers have over the years been linked to poor programming practices that lead to potential resource leaks, memory corruption, and security-related bugs. Because pointers allow indiscriminate access to the computer memory, it is relatively easy to misuse them, resulting in bugs that are difficult to fix.

Some C++ classes and templates provide a great alternative to pointers, with little overhead and many of the same features. The main technique to avoid traditional pointers is to use smart pointer templates. Such smart pointers are simple templates that can be used to store addresses in a safer way. For example, a smart pointer of the type std::shared_ptr is a template-type object that allows the same address to be used by two or more parts of the code.

The main advantage of a smart pointer over a traditional C++ pointer is that the smart pointer knows how to clean up itself when it is no longer needed. Thus, a large class of problems that occur when a programmer doesn't dispose of the pointer is avoided. The cleanup mechanism is defined according to the rules of the RAII (Resource Acquisition Is Initialization) idiom: resources contained in a smart pointer are initialized in the constructor of the template object, and released during destruction, which typically happens when the object in question goes out of scope.

There are different types of smart pointers, each one designed for a particular use case. The most commonly used smart pointers in C++ code are unique pointers and shared pointers.

A **unique pointer** (of template type `unique_ptr`) provides a wrapper for a traditional C++ pointer. The template, however, defines the semantic of object ownership, so that other references to the pointer are not valid after the transfer of ownership occurs. A unique pointer can be used in situations in which the receiver will take full control of the pointed object, as well as any associated resource. Therefore, a pointer passed to a `unique_ptr` object should not be referenced again in contexts other than the one where the `unique_ptr` is used.

A **shared pointer** (of template type `shared_ptr`) is a template that can be used to wrap an existing C++ native pointer. Unlike a unique pointer, a shared pointer can be used by two or more parts of the code. Internally, shared pointers maintain a counter that can be used to determine how many references to the original pointer have been created. This type of mechanism is referred to as reference counting. Every time a shared pointer object is destroyed, it checks its internal reference counter. If the counter is greater than zero, the internal object is not deleted. However, if the reference counter reaches zero, then the referred object is deleted and its destructor is activated.

In this section, you will see the details of using a unique pointer. In the next section, I will introduce shared pointers.

Using Unique Pointers

The solution to our memory management problem involves the use of unique pointers. A unique pointer is implemented using the standard class `std::unique_ptr` and, like other smart pointers, can be used to provide automated cleanup and resource ownership.

The policy used by `std::unique_ptr` ensures that once it has been assigned the ownership of the pointer, the memory stored in the pointer is not owned by anyone else.

Therefore, the semantic of an `std::unique_ptr` involves the automatic destruction of the associated data as soon as the object goes out of scope. If the owner of the unique pointer doesn't want to destroy the memory, it has the option of moving it to another unique pointer, in a process of transferring the ownership to another object. This is how I solved the issue of managing the memory associated to the returned `FinancialStatement`.

First, consider the definition of the `FinancialStatement` class.

```
class FinancialStatement {
public:
    FinancialStatement();
    ~FinancialStatement();
    FinancialStatement(const FinancialStatement&);
    FinancialStatement &operator=(FinancialStatement &);

    double getReturn();
    void addTransaction(const std::string &security, double val);

private:
    double m_return;
    std::vector<std::pair<std::string,double>> m_transactions;
};
```

There is also a function that is used to create a sample financial statement.

```
std::unique_ptr<FinancialStatement> getSampleStatement();
```

This function returns a unique pointer to a sample statement. This means that the caller of the code owns the returned memory, since the caller will have a unique pointer that has been initialized with a pointer to the resulting object. The implementation of getSampleStatement is

```
std::unique_ptr<FinancialStatement> getSampleStatement()
{
    std::unique_ptr<FinancialStatement> fs(new FinancialStatement);
    fs->addTransaction("IBM", 102.2);
    fs->addTransaction("AAPL", 523.0);
    return fs;
}
```

After the `FinancialStatement` object has been allocated and used to create a unique pointer, it is initialized and finally returned to the caller. Since the return statement transfers the ownership of the pointer to the returned object, the original `FinancialStatement` object is not destructed. Instead, it is now owned by the caller of the `getSampleStatement` function.

Complete Code

Listing 3-2 presents the code for the `FinancialStatement` class, which is divided into a header file and an implementation file.

Listing 3-2. FinancialStatement.h

```
//
//  FinancialStatement.h

#ifndef __FinancialSamples__FinancialStatement__
#define __FinancialSamples__FinancialStatement__

#include <string>
#include <vector>

class FinancialStatement {
public:
    FinancialStatement();
    ~FinancialStatement();
    FinancialStatement(const FinancialStatement&);
    FinancialStatement &operator=(FinancialStatement &);

    double getReturn();
    void addTransaction(const std::string &security, double val);

private:
    double m_return;
    std::vector<std::pair<std::string,double> > m_transactions;
};

std::unique_ptr<FinancialStatement> getSampleStatement();
```

```cpp
void transferFinancialStatement(std::unique_ptr<FinancialStatement>
&statement);

#endif /* defined(__FinancialSamples__FinancialStatement__) */

//
//  FinancialStatement.cpp

#include "FinancialStatement.h"

FinancialStatement::FinancialStatement()
: m_return(0)
{

}

FinancialStatement::~FinancialStatement()
{

}

FinancialStatement::FinancialStatement(const FinancialStatement &v)
: m_return(v.m_return),
m_transactions(v.m_transactions)
{

}

FinancialStatement &FinancialStatement::operator=(FinancialStatement &v)
{
    if (this != &v)
    {
        m_return = v.m_return;
        m_transactions = v.m_transactions;
    }
    return *this;
}
```

```cpp
double FinancialStatement::getReturn()
{
    return m_return;
}

void FinancialStatement::addTransaction(const std::string &security,
double val)
{
    m_transactions.push_back(std::make_pair(security, val));
}

// returns a sample statement that includes a few common stocks
std::unique_ptr<FinancialStatement> getSampleStatement()
{
    std::unique_ptr<FinancialStatement> fs(new FinancialStatement);
    fs->addTransaction("IBM", 102.2);
    fs->addTransaction("AAPL", 523.0);
    return fs;
}

void transferFinancialStatement(std::unique_ptr<FinancialStatement>
statement)
{
    // perform transfer here
    // ...
    // statement is still valid
    std::cout << statement->getReturn() << std::endl;
    // statement is released here
}

//
//  main.cpp

#include "FinancialStatement.h"

#include <iostream>
```

```
int main()
{
    std::unique_ptr<FinancialStatement> fs = getSampleStatement();
    // do some real work here...
    return 0;
    // the unique pointer is released at the end of the scope...
}
```

Transferring Ownership

In the previous example, you saw how to use unique pointers to transfer the ownership of objects to the caller of a function. Another important use of these smart pointers is to tell to a caller that the called function is taking ownership of the passed object.

For example, consider the function transferFinancialStatement, which is defined in the following way:

```
void transferFinancialStatement(std::unique_ptr<FinancialStatement>
statement);
```

The parameter statement is a unique pointer to a FinancialStatement object, which means that once it receives a parameter of that particular type, it will become the owner of the object. Thus, the transferFinancialStatement can use the statement object knowing that it is the sole owner of its contents. Depending on the operations necessary to perform on the object, this may be an important advantage.

The sample implementation of the function reads as follows:

```
void transferFinancialStatement(std::unique_ptr<FinancialStatement>
&statement)
{
    // perform transfer here
    // ...
    // statement is still valid
    std::unique_ptr<FinancialStatement> another = std::move(statement);
    std::cout << statement->getReturn() << std::endl;
    // statement is released here
}
```

The important thing to understand here is that the pointed object is destroyed at the end of the `transferFinancialStatement`, since the parameter was transferred and therefore goes out of scope at the end of the function block.

Pitfalls of Unique Pointers

Due to the semantics of unique pointers, which require the ownership of the referenced data, a few errors can occur when programmers try to access data stored in this kind of smart pointer. For example, a common source of errors is the use of functions that take ownership of the pointed object. You can see this class of errors by using the function `transferFinancialStatement`, which I showed previously. When a unique pointer object is passed to a function that takes ownership of the memory, the unique pointer is not valid anymore, and may cause a crash the next time it is accessed.

```
int main()
{
    std::unique_ptr<FinancialStatement> fs = getSampleStatement();
    transferFinancialStatement(fs);

    // the unique_ptr object is invalid here, the next access can crash the
    program
    std::cout << fs->getReturn() << "\n";

    return 0;
}
```

This example shows how easy it is to crash a program after a unique pointer has transferred the ownership of the object it points to. This is done as you call the `transferFinancialStatement` function.

The next line tries to access the return value, which generates an invalid access. This results in a segmentation fault in many platforms. To avoid such invalid accesses, a programmer needs to be careful about calling functions that accept a unique pointer, vs. a native pointer.

Another pitfall of unique pointers is their lack of support for STL containers. For example, it is not possible to have unique pointers as members of an `std::vector`. This happens because most containers work by copying their elements by value, which in practice invalidates the data stored in the existing unique pointer. Moreover, many algorithms for STL containers perform internal copies of the data they contain. This means that such algorithms can destroy the original elements without previous notice: a very awkward situation. For these reasons, it is recommended that you avoid using unique pointers when dealing with STL containers.

To avoid the problems with `std::unique_ptr`, another smart pointer type has been introduced: the `std::shared_ptr` template provides the semantics of pointers that can be shared by different owners. In the next section, you will see an example of how to use such shared pointer objects.

Determining Credit Ratings

Create a class to determine the credit ratings of a given security.

Solution

Credit risk ratings are defined by accredited rating agencies and used to determine the risk of institutions. With this information, investors can determine the risk level of a particular bond or stock holding and control the level of risk they are willing to assume. For example, risk-averse investors such as pension funds and insurance companies typically shy away from investing in any institution that is not certified as top grade (typically AAA). There are three main credit rating agencies used by financial institutions all over the world: Moody's, S&P, and Fitch. They define the risk grade not only for companies but also for local-, state-, and national-level governments.

To model credit ratings in C++, I created a separate class that encapsulates the fundamental concepts behind credit risk rating. The class `CreditRisk` has a member function called `getRating`, which returns the prevailing risk rating for a particular stock or bond. A second class, named `RiskCalculator`, is used to perform a simple analysis of the risk associated with a particular portfolio, given a set of risk ratings for each component of the portfolio.

Using Shared Pointers

Since the `RiskCalculator` class needs to maintain a set of credit ratings, it makes sense to have a container to hold this information. However, we would like to avoid making copies of the data. Since the `CreditRisk` class is so simple, it doesn't make much difference if you make copies or not. However, classes can become more complex, and in a large application, these memory requirements start to add up. Therefore, a better design is to avoid making copies of CreditRisk objects as we add them to collections. We have seen, however, that std::unique_ptr is not suitable for inclusion in collections, which leaves us with the requirement of storing traditional pointers for objects.

A better solution is to use the std::shared_ptr class to handle the memory associated with CreditRisk objects. A shared pointer not only knows how to clean up the data referenced by a pointer but is also able to share the reference with other objects. This way, the object will not be destroyed until the last reference is also destroyed.

Shared pointers achieve this behavior through the use of a reference counting mechanism. A counter is maintained by the shared pointer object, which determines how many copies exist for the referenced object. When the shared pointer is destroyed, it checks this counter to determine if other references exist. If the counter is positive, the pointed object is not destroyed. The counter is also updated when a new copy of the shared pointer is created. The counter is incremented to indicate that another copy of the object reference exists. In this way, several copies of the same shared pointer can live in memory, and they will all manage the underlying data correctly, until the last one is destroyed and the original object is removed from memory.

I used shared pointers to handle the memory requirements of the RiskCalculator class. The set of CreditRisk objects is maintained as a vector of shared pointers, declared as follows:

```
std::vector<std::shared_ptr<CreditRisk> > m_creditRisks;
```

In this declaration, each element of the vector is a shared pointer. Since shared pointers know how to copy themselves, they can be used effectively as members of a vector or any other container, unlike unique pointers.

Working with shared pointers is easy because they automatically perform the required cleanup actions. For example, copying unique pointers is done simply with the use of the assignment operator. New elements to the m_creditRisks vector are added by the addCreditRisk member function.

```
void RiskCalculator::addCreditRisk(std::shared_ptr<CreditRisk> risk)
{
    m_creditRisks.push_back(risk);
}
```

Complete Code

In Listing 3-3, you have the complete code for the example described in the previous section. The class is called `CreditRisk`, which is contained in a header and an implementation file.

Listing 3-3. CreditRisk.h

```
//
// CreditRisk.h

#ifndef __FinancialSamples__CreditRisk__
#define __FinancialSamples__CreditRisk__

// A simple class representing a credit risk assessment
class CreditRisk {
public:
    // these are risk grades, as determined by rating agencies
    enum RiskType {
        AAA,
        AAPlus,
        AA,
        APlus,
        A,
        BPlus,
        B,
        CPlus,
        C
    };

    // other methods here ...
};

#endif /* defined(__FinancialSamples__CreditRisk__) */

//
// RiskCalculator.h
```

```cpp
#ifndef __FinancialSamples__RiskCalculator__
#define __FinancialSamples__RiskCalculator__

#include "CreditRisk.h"

#include <memory>
#include <vector>

// calculates the risk associated to a portfolio
class RiskCalculator {
public:
    RiskCalculator();
    ~RiskCalculator();
    RiskCalculator(const RiskCalculator &);
    RiskCalculator &operator =(const RiskCalculator &);

    void addCreditRisk(std::shared_ptr<CreditRisk> risk);

    CreditRisk::RiskType portfolioMaxRisk();
    CreditRisk::RiskType portfolioMinRisk();
private:
    std::vector<std::shared_ptr<CreditRisk> > m_creditRisks;
};

#endif /* defined(__FinancialSamples__RiskCalculator__) */

//
//  RiskCalculator.cpp

#include "RiskCalculator.h"

RiskCalculator::RiskCalculator()
{

}

RiskCalculator::~RiskCalculator()
{

}
```

```cpp
RiskCalculator::RiskCalculator(const RiskCalculator &v)
: m_creditRisks(v.m_creditRisks)
{

}

RiskCalculator &RiskCalculator::operator =(const RiskCalculator &v)
{
    if (this != &v)
    {
        m_creditRisks = v.m_creditRisks;
    }
    return *this;
}

void RiskCalculator::addCreditRisk(std::shared_ptr<CreditRisk> risk)
{
    m_creditRisks.push_back(risk);
}

CreditRisk::RiskType RiskCalculator::portfolioMaxRisk()
{
    CreditRisk::RiskType risk = CreditRisk::RiskType::AAA;

    for (int i=0; i<m_creditRisks.size(); ++i)
    {
        if (m_creditRisks[i]->getRating() < risk)
        {
            risk = m_creditRisks[i]->getRating();
        }
    }
    return risk;
}

CreditRisk::RiskType RiskCalculator::portfolioMinRisk()
{
    CreditRisk::RiskType risk = CreditRisk::RiskType::C;
```

```
    for (int i=0; i<m_creditRisks.size(); ++i)
    {
        if (m_creditRisks[i]->getRating() > risk)
        {
            risk = m_creditRisks[i]->getRating();
        }
    }
    return risk;
}
```

Using the auto Keyword

Among the many additions to C++ introduced by the C++11 standard, the auto
keyword is one of the most practical and easy to understand. The basic idea behind this
extension is that the compiler can perform the job of type detection for many categories
of variables and expressions. Whenever this is possible, the programmer can use the
auto keyword in the variable declaration, instead of entering the full type of the desired
object. This way, the programmer can decide to let the compiler do the type detection
automatically while using the type name only when a visual labeling is desired and
convenient.

The auto keyword has a few advantages over other forms of type declaration. It

- Reduces the amount of manual work done by programmers, since it
  uses the compiler itself to analyze an expression and determine the
  exact type for a particular variable or expression

- Adds uniformity to the code, as most variables are declared using the
  same style

- Leaves intact the ability of programmers to enter the exact type as
  desired, so that any conflicts can be avoided

- Simplifies the use of templates, because data types can be
  automatically detected during compilation at each instantiation of
  the template

As an example of the use of the auto keyword, we can modify some of the member
functions in the RiskCalculator class to perform automatic detection of variable types.

This is useful to simplify some of the expressions that are so common when using template collections. The following, for instance, is the portfolioMaxRisk member function:

```
CreditRisk::RiskType RiskCalculator::portfolioMaxRisk()
{
auto risk = CreditRisk::RiskType::AAA;
for (auto &p : m_creditRisks)
{
    if ((*p)->getRating() < risk)
    {
        risk = (*p)->getRating();
    }
}
return risk;
}
```

This code snippet shows how the type of the local variable risk can be automatically detected, so you don't need to enter the redundant name of the type CreditRisk::RiskType. Similarly, the next line shows how to iterate through a collection of objects using the auto keyword to determine the right type of the iterator. Seasoned STL programmers know that using iterators may introduce a lot of extra types to a piece of code, which frequently obfuscates the original intent of the program. For comparison, notice that if it weren't for the auto keyword, the foregoing for loop would need to be rendered as

```
for (std::vector<std::shared_ptr<CreditRisk> >::iterator p = m_creditRisks.
begin();
    p != m_creditRisks.end(); ++p)
{
    // ...
}
```

Not only is this harder to enter manually, but it also makes the code harder to understand and modify.

Collecting Transaction Data

Create a solution to the problem of handling transaction orders, including BUY, SELL, or SELL SHORT, stored in a single file. The solution must correctly handle programming exceptions.

Solution

To solve this problem, we created a class that can handle trading transactions and perform the necessary operations. The class is responsible for receiving a filename and executing the instructions stored in the file. The class is also responsible for handling any error happening during this process, including errors from reading the file as well as incorrect trading requests sent to the application.

In this coding example, the operations allowed are simple and include only buy, sell, and sell short. Therefore, I concentrate on the problem of processing the file and handling unexpected errors in the program. We follow the best practices of using the exception handling mechanism offered by C++. Therefore, we create a new class, called `TransactionHandler`, which is able to read data from a file and perform the necessary actions in the member function `handleTransactions`. The resulting code is able to execute the trading actions stored in the file, but it handles possible exceptions using the `try/catch/throw` mechanism supplied by the C++ language as described in the next section.

Exception Handling

One of the basic problems faced by programmers is detecting and recovering from errors. While we try to avoid self-inflicted errors, there are many extraordinary situations that need to be handled even by correct programs. For example, what should your code do when the file system is full and there is no more space to save the current file? What can be done when a network connection is closed and the server is not available to complete a download? Programmers need to decide on how to respond to such exceptional situations, and a few strategies have been devised over the years in order to respond to such conditions.

C++ uses an exception-based model to deal with unexpected conditions occurring during program execution. This model uses a standard `try/catch` block to contain the code that you may want to protect. When an exception happens, the compiler throws an object to indicate the unexpected condition. As a result, the enclosing blocks of code will also destroy all local objects that have been created in that particular context, to avoid resource leaks.

The other aspect of exception handling in C++ is the use of exception objects that inform programmers about the class of error that occurred. These objects are created using the throw keyword and caught using the catch block, which receives a reference to the exception object and uses it to understand and possibly recover from an unexpected state. Applications are free to create new classes of exceptions as a way to provide additional information about the error that triggered the exception.

In this example, I created two exception classes. They both derive from std::runtime_error, which provides the basic behavior for runtime exceptions. The first class, FileError, is used to flag any error occurred during the process of reading the file. The following is its definition:

```cpp
class FileError :public std::runtime_error {
public:
    FileError (const std::string &description);
};
```

The second exception class, TransactionTypeError, is thrown when an unknown transaction type is found in the file, other than TRANSACTION_SELL, TRANSACTION_BUY, or TRANSACTION_SHORT. The definition is similar to what you saw with FileError. These classes are then used on the test code to determine the type of error encountered and how to proceed. In the test application, we just print a descriptive error message using the string returned by the what() member function before terminating the program.

```cpp
try
{
    TransactionHandler handler(fileName);
}
catch (FileError &e)
{
    std::cerr << "received a file error: " << e.what() << std::endl;
}
```

```
catch (TransactionTypeError &e)
{
    std::cerr << "received a transaction error: " << e.what() << std::endl;
}
catch (...)
{
    std::cerr << "received an unknown error\n";
}
```

Complete Code

Listing 3-4 provides a complete example of exception handling. The classes presented here demonstrate how to handle exceptions while reading a transaction file. A sample main() function, provided at the end, shows how these classes work together.

Listing 3-4. TransactionHandler.h

```
//
// TransactionHandler.h

#ifndef __FinancialSamples__TransactionHandler__
#define __FinancialSamples__TransactionHandler__

#include <iostream>

enum TransactionType {
    TRANSACTION_SELL,
    TRANSACTION_BUY,
    TRANSACTION_SHORT,
};

class FileError :public std::runtime_error {
public:
    FileError(const std::string &s);
};
```

```cpp
class TransactionTypeError :public std::runtime_error {
public:
    TransactionTypeError(const std::string &s);
};

class TransactionHandler {
public:
    static const std::string SELL_OP;
    static const std::string BUY_OP;
    static const std::string SHORT_OP;

    TransactionHandler(const std::string &fileName);
    TransactionHandler(const TransactionHandler &);
    ~TransactionHandler();
    TransactionHandler &operator=(const TransactionHandler&);

    void handleTransactions();
private:
    std::string m_fileName;
};

#endif /* defined(__FinancialSamples__TransactionHandler__) */

//
//  TransactionHandler.cpp

#include "TransactionHandler.h"

#include <fstream>

FileError::FileError(const std::string &s)
: std::runtime_error(s)
{
}

TransactionTypeError::TransactionTypeError(const std::string &s)
: std::runtime_error(s)
{

}
```

```cpp
const std::string TransactionHandler::SELL_OP = "SELL";
const std::string TransactionHandler::BUY_OP = "BUY";
const std::string TransactionHandler::SHORT_OP = "SHORT";

TransactionHandler::TransactionHandler(const std::string &fileName)
: m_fileName(fileName)
{

}

TransactionHandler::TransactionHandler(const TransactionHandler &a)
: m_fileName(a.m_fileName)
{

}

TransactionHandler::~TransactionHandler()
{
}

TransactionHandler &TransactionHandler::operator=(const
TransactionHandler&a)
{
    if (this != &a)
    {
        m_fileName = a.m_fileName;
    }
    return *this;
}

void TransactionHandler::handleTransactions()
{
    std::ifstream file;
    file.open(m_fileName, std::ifstream::in);
    if (file.fail())
    {
        throw new FileError(std::string("error opening file ") +
        m_fileName);
    }
```

```cpp
    std::string op;
    file >> op;
    while (file.good() && !file.eof())
    {
        if (op != SELL_OP && op != BUY_OP && op !=  SHORT_OP)
        {
            throw new TransactionTypeError(std::string("unknown
            transaction ") + op);
        }

        // process remaining transaction data...
    }
}

//
//  main.cpp

#include "TransactionHandler.h"

#include <iostream>

int main(int argc, const char **argv)
{
    if (argc < 2)
    {
        std::cerr << "usage: <progName> <fileName>\n";
        return 1;
    }

    std::string fileName = argv[1];
    try
    {
        TransactionHandler handler(fileName);
    }
    catch (FileError &e)
    {
        std::cerr << "received a file error: " << e.what() << std::endl;
    }
```

```
    catch (TransactionTypeError &e)
    {
        std::cerr << "received a transaction error: "
                    << e.what() << std::endl;
    }
    catch (...)
    {
        std::cerr << "received an unknown error\n";
    }
    return 0;
}
```

Implementing Vector Operations

In this section, we implement the common addition and multiplication operators defined on numerical vectors.

Solution

This problem can be easily solved using the C++ facilities for operator overloading. I first give a general introduction to this programming technique and subsequently show how to use it to implement numerical vector operations.

Operator Overloading

Operators are used in most programming languages to provide a simpler syntax for common operations. For example, the + operator is used to implement the addition of numbers without the need for a function called sum. So, one can type the following expression:

```
int total = a + b + c + d;
```

instead of the less convenient version

```
int total = sum(a, sum(b, sum(c, d)));
```

Similarly, other operators perform comparable tasks for other primitive operations, such as subtraction, multiplication, logical comparison, and pointer arithmetic.

While operators are available in most modern programming languages, C++ is one of the few languages that allow programmers to redefine the meaning of existing operators to adapt them to the most natural usage in the target domain. For example, in an application where vectors of numbers are a common data structure, it makes sense to redefine the + operator to perform the sum of vectors in addition to the traditional usage of adding numeric (scalar) values.

C++ allows the definition of operators for each new declared type. Operators can be defined as part of a class (as a member function) or as a freestanding function. For example, consider the class Complex that redefines the + operator. The declaration for the operator can be written as

```
class complex {
public:
    // ... other methods here
    complex &operator +(const complex &v);
};
```

Another way of writing the same operator is using a free function.

```
Complex &operator +(const complex &a, const complex &b);
```

The difference between these two declarations is that the latter declares a function that receives two parameters, while the member function version requires only one additional parameter (the first parameter is the object itself). Similarly, you can declare new versions of most C++ operators, including math, logical, and pointer operators.

To solve the problem posed, we created a new class called NumVector, a simple numerical vector that can be used to store double numbers. To provide operations that can be applied to a vector in a natural way, we use operators that are declared as free functions.

```
NumVector operator +(const NumVector &a, const NumVector &b);
NumVector operator -(const NumVector &a, const NumVector &b);
NumVector operator *(const NumVector &a, const NumVector &b);
```

The class also provides a few member functions that are used in the implementation of these operations. In particular, you will want to have the member functions add, which adds a new element to the vector, removeLast, which removes the last element, get, which returns one of the elements given a position (index), and finally a size member function, which returns the size of the vector.

All operators are implemented in a similar way: they all check that the two parameters have the same size and then perform a loop that is used to perform the required operation—addition, subtraction, or multiplication.

Complete Code

Listing 3-5 provides a sample implementation of a numeric vector. You can use this class to create numerical vectors and perform common vector operations.

Listing 3-5. NumVector.h

```
//
//  NumVector.h

#ifndef __FinancialSamples__NumVector__
#define __FinancialSamples__NumVector__

#include <vector>

class NumVector {
public:
    NumVector();
    ~NumVector();
    NumVector(const NumVector &);
    NumVector &operator =(const NumVector &);

    void add(double val);
    void removeLast();
    double get(int pos) const;
    size_t size() const;
private:
    std::vector<double> m_values;
};
```

```cpp
NumVector operator +(const NumVector &a, const NumVector &b);
NumVector operator -(const NumVector &a, const NumVector &b);
NumVector operator *(const NumVector &a, const NumVector &b);

#endif /* defined(__FinancialSamples__NumVector__) */

//
//  NumVector.cpp

#include "NumVector.h"

#include <iostream>

NumVector::NumVector()
{

}

NumVector::~NumVector()
{

}

NumVector::NumVector(const NumVector &v)
: m_values(v.m_values)
{

}

NumVector &NumVector::operator=(const NumVector &v)
{
    if (this != &v)
    {
        m_values = v.m_values;
    }
    return *this;
}

size_t NumVector::size() const
{
    return m_values.size();
}
```

```cpp
double NumVector::get(int pos) const
{
    return m_values[pos];
}

void NumVector::add(double val)
{
    m_values.push_back(val);
}

void NumVector::removeLast()
{
    m_values.pop_back();
}

NumVector operator +(const NumVector &a, const NumVector &b)
{
    if (a.size() != b.size())
    {
        throw new std::runtime_error("vectors must have the same size");
    }
    NumVector result;
    for (int i=0; i<a.size(); ++i)
    {
        result.add(a.get(i) + b.get(i));
    }
    return result;
}

NumVector operator -(const NumVector &a, const NumVector &b)
{
    if (a.size() != b.size())
    {
        throw new std::runtime_error("vectors must have the same size");
    }
    NumVector result;
```

```
    for (int i=0; i<a.size(); ++i)
    {
        result.add(a.get(i) - b.get(i));
    }
    return result;
}

NumVector operator *(const NumVector &a, const NumVector &b)
{
    if (a.size() != b.size())
    {
        throw new std::runtime_error("vectors must have the same size");
    }
    NumVector result;
    for (int i=0; i<a.size(); ++i)
    {
        result.add(a.get(i) * b.get(i));
    }
    return result;
}
```

Conclusion

C++ is a complex language, which provides mechanisms for the creation of software using one or more among several paradigms, such as structured, object-oriented, and functional programming. As a result, it is necessary to develop a set of techniques that are more appropriate for the development of financial software while avoiding unproductive practices that obfuscate programs and hinder our ability to modify them. Over the years, financial engineers have successfully used a number of C++ idioms that make it easier to use the speed and abstraction facilities of the language.

In this chapter, I discussed a few programming examples that introduced and reviewed some of these useful C++ programming techniques, which are commonly employed in the development of financial applications. First, I reviewed the concept of templates, which let programmers write generic code that can be applied to several classes. In the first code sample, you learned how to design an interest rate calculation engine that is independent of the definition of interest rate classes. This type of design is very useful in the creation of large-scale financial applications.

Next, you learned how to define financial statement objects that can be sent to other parts of the application while reducing the occurrence of memory leaks. For this purpose, I explained the use of smart pointers as a way of making automatic decisions about the lifetime of objects. In particular, you learned about the `std::unique_ptr` template, which implements the semantics of auto-released, self-owned pointers.

The next section dealt again with memory management issues, this time in the context of determining credit ratings from rating agencies. In this case, you learned how to share rating information in such a way that the memory would be automatically destroyed even when several users had copies of the object. For this purpose, you can learn about the `std::shared_ptr` template, which uses a reference counting mechanism to determine the correct moment to delete memory, therefore avoiding memory leaks. You have also seen the use of the `auto` keyword to simplify type detection with STL containers. Both `shared_ptr` and the `auto` keyword are new features introduced with the C++11 standard, which is currently implemented by all modern C++ compilers.

Another important technique in C++ involves the handling of unexpected conditions. The exception-based mechanism provided by C++ allows programmers to deal with infrequent conditions in a clean way. You have seen an example of such policies applied to the problem of processing trading operations stored in a data file.

Finally, this chapter has considered the problem of implementing mathematical operations applied to a sequence of numbers. For this purpose, we created a new class called `NumVector`, which stores numbers in a sequence. To implement the addition, subtraction, and multiplication of vectors in a natural way, we used the operator overloading mechanism provided by C++. In this way, an application can perform vector operations using operators already present in the language, redefined so that they can be applied to your own types.

In this chapter, you learned about general C++ techniques that have been successfully used in financial applications. In the next chapter, I explain how to use libraries that extend the language and offer useful functionality for financial software developers. These libraries include facilities such as new data containers and algorithms, as well as more advanced techniques for memory management, as well as time and event handling.

CHAPTER 4

Common Libraries for Financial Applications

Financial code implemented in C++ uses programming libraries designed to simplify the creation of fast, standard-conformant classes. The best example of such libraries is the STL (standard template library) itself, a convenient library that is included with standard-compliant C++ compilers. The STL offers a set of generic, commonly used containers that may be applied to almost any situation. Knowing how to employ well the STL is one of the main skills necessary for effective C++ programming, especially in the context of high-performance software development—a common requirement for financial applications. In this chapter, you will learn programming examples that clarify some of the most common uses of the STL for financial programming, including containers and algorithms.

The boost project provides another set of commonly used classes. Although the standard language committee does not officially support boost libraries, some of them have been used as the basis for additions to the last few versions of the C++ standard library. Therefore, a good understanding of the classes and templates included in the boost repository is valuable as a way to have early access to functionality that will only later be made available in all C++ implementations.

In the next few sections, we explore C++ examples that illustrate how these classes and templates are used in financial applications. Examples of important library components explored here include

- Vectors: These are containers used to manipulate objects of the same type.

- Maps: STL containers that can be used to associate values to a set of keys, which can be of any type.

113

© Carlos Oliveira 2021
C. Oliveira, *Practical C++20 Financial Programming*, https://doi.org/10.1007/978-1-4842-6834-6_4

- Algorithms: The STL also provides a rich set of algorithms that can be used to manipulate the standard containers. You can also extend the existing algorithms so that they can be applied to your own data structures. Similarly, you can use the ideas provided by STL to implement your own algorithms.

- Boost libraries: The STL is the foundation for other useful libraries. Boost libraries are written by some of the C++ experts working on the language committee. Many of the components previously included in the boost library, such as `shared_ptr`, have since become part of the language.

- Time and date handling: Financial applications are usually related to the processing of prices over specific time periods. To make this possible, it is necessary to use libraries to handle date uniformly.

In the next sections, you will see a few selected programming examples that explore some of these C++ libraries in the context of financial applications.

Handling Analyst Recommendations

One of the common events around a particular stock is the release of analyst recommendations. Create a C++ class that handles analyst recommendations and returns the average target price for the stock.

Solution

Analyst recommendations are an important part of the Wall Street ecosystem. Many financial institutions such as pension funds and insurance companies, as well as retail investors, use analysts' recommendations as a gauge of the predominant view about a particular stock. This in turn can be used to determine future capital allocation to a portfolio of stocks.

Analyst recommendations come from one of the several institutions that provide public analysis of equity investments, generally from some of the major investment banks. The recommendation for a particular stock includes a defined action such as "buy," "sell," or "hold." The recommendation also frequently includes a price target, which determines how much the analyst expects to be the "fair price" for the instrument.

Since there are so many analysts covering the equities market, keeping track of recommendations is one of the important parts of the work of an institutional investor. In this section, you will create a C++ class to store this type of information and to answer some basic questions such as "what is the average target price for a particular stock?"

The solution for this problem involves the use of STL containers to hold the data. In particular, you will use vectors to provide quick access to the data.

More About STL Vectors and Maps

The STL is a repository of standard data structures and algorithms that are useful in most programming domains. For financial applications in particular, the use of the STL is extremely important because STL components are optimized for high speed. For example, STL components such as vectors are currently the preferred way to write applications in C++, instead of using raw C++ arrays and similar data-container classes. Compiler vendors have done a great job of making STL components fast and safe to use in a wide range of applications, so that programmers don't need to worry about intricate issues such as memory allocation, exception handling, and algorithm complexity.

STL vectors are versatile because they can grow dynamically. Therefore, they work in situations where you don't have a clear idea of how many elements will be stored in the underlying data structure (and as long as you don't care about the overhead of vector resizing). For example, if you're reading trading data over a given time period, you typically don't know how many trades occurred during that particular time frame. In this situation, it is easier to use a vector that can be initialized with a small number of elements and then grow as needed, instead of using an array with a predefined (and fixed) size. By using the STL vector in this way, you don't need to worry about memory allocation and exception safety of the container.

The vector template exposes an interface with operations that can be applied to a set of elements, such as adding, removing, finding, and comparing. These common operations can be used across concrete implementations, without any manual changes required. For example, vectors declared as `std::vector<int>` can use the `push_back` member function to add `int` elements to the end of the vector. Similarly, a second vector declared as `std::vector<std::string>` can also add `std::string` elements using the same template-based member function.

The following is a quick list of the most regularly used member functions in the `std::vector` template:

- `template <class T> vector(const T &c, int n)`: Constructor used to create a new vector initialized with n copies of the constant element c.

- `template <class T> vector(const vector<T> &v)`: Constructor used to create a new vector initialized with a copy of an existing vector object v.

- `template <class T> T &operator [] (int pos)`: This operator makes the `std::vector` object behave as a native array. You can use the notation v[i] to access or update the value of an element stored at position i of vector v.

- `template <class T> void push_back(const T &c)`: Used to add a new element to the back of a vector, allocating additional memory if necessary. This is the most commonly used way to add new elements to a vector, since it takes care of adding memory to store the new element when necessary, unlike the `operator []`, which will crash the application when an undefined position is accessed.

- `template <class T> void pop_back()`: This member function performs the inverse of the push_back operation, removing the element stored in the last position of the vector. However, the memory allocated for that element is not immediately reclaimed, and it may be used by later operations.

- `template <class T> size_t size()`: This member function returns the number of elements currently stored in the `std::vector`. Notice that this may be less than the total memory currently used by the vector, since it is possible for the vector to allocate more memory than it currently needs, depending on the number of elements previously added or reserved.

In our problem, we use vectors to store recommendations for a particular stock. Each stock covered by this class will have a vector of recommendations. Each recommendation is just an object of the `Recommendation` class, which stores recommendations defined in the following way:

```
enum class RecommendationType {
    BUY,
    SELL,
    HOLD,
    NO_RECOMMENDATION
};
```

The Recommendation class is defined as

```
class Recommendation {
public:
    Recommendation();
    Recommendation(const std::string &ticker, RecommendationType rec,
    double target);
    ~Recommendation();
    Recommendatioperatoron(const Recommendation &r);
    Recommendation & =(const Recommendation &r);

    double getTarget() const;
    RecommendationType getRecommendation() const;
    std::string getTicker() const;

    // private members
};
```

This simple Recommendation class stores the ticker for stock, as well as its recommendation type and price target. Notice that objects that are stored in a std::vector need to be from classes that can be copied or moved, since elements in a std::vector are stored by value. This means that a copy is created whenever there is the need to move the object to a certain position, unless the class has a movable constr.

Using std::vector we can keep track of the individual recommendation for a stock. However, there are several stocks in the universe of equities that we would like to track. To find the right recommendations, we assign a way to retrieve objects based on the stock ticker. This is performed using a std::map template. Using a map will also simplify the code necessary to implement other useful operations, such as adding new recommendations or calculating the average recommendation.

The std::map template provides a way to associate an arbitrary key to a data object, so that you can easily retrieve the original data. The best thing about maps is that the key can be of any kind of object offering a comparison operation, such as the less than (<) operator. In our case, for example, you can use a string that indicates the unique ticker for each particular stock. Based on the ticker, you can retrieve the vector that contains the recommendations for that particular stock.

Here is a short list of important operations defined for `std::map`:

> `template <class K, class T> iterator<T> find(const K&):`
> Returns an iterator to the data object that is associated to the given key. If the key has no association, the function returns the end() iterator.

> `template <class K, class T> T &operator[](const K&):`
> Associates a key with a particular data object. You can use this operator to retrieve elements from the map container as well as insert new elements.

> `template <class K, class T> size_t erase(const K&):`
> Erases the data associated with the given key.

The `std::map` template is used in this code example to store and retrieve recommendations that were issued for a particular stock. The `RecommendationProcessor` is the class responsible for storing, processing, and answering queries related to stock recommendations. The general algorithm used by `RecommendationProcessor` consists of storing new recommendations in an internal data structure. Then, at any future moment, you can query the stored data using the `averageTargetPrice` member function.

The first member function in this class, `addRecommendation`, is responsible for storing new recommendations to the `m_recommendations` member variable.

```
void RecommendationsProcessor::addRecommendation(const std::string &ticker,
                                                 RecommendationType rec,
                                                 double targetPrice)
{
    Recommendation r(ticker, rec, targetPrice);
    m_recommendations[ticker].push_back(r);
}
```

Here, you first need to create a new recommendation object based on the information passed, such as ticker, type of recommendation, and target price. Then, we use the m_recommendations map to access the vector of recommendations. Finally, we use the push_back method to add the new recommendation to the list of recommendations for that particular stock.

Another interesting member function is the one that calculates the average target price. It looks at all recommendations to calculate an average target.

```
double RecommendationsProcessor::averageTargetPrice(const std::string
&ticker)
{
    if (m_recommendations.find(ticker) == m_recommendations.end())
        return 0;
    auto vrec = m_recommendations[ticker];
    std::vector<double> prices;
    for (auto i=0; i<vrec.size(); ++i)
    {
        prices.push_back(vrec[i].getTarget());
    }
    return std::accumulate(prices.begin(), prices.end(), 0) / prices.size();
}
```

The first thing you need to do in this code is to check if the stock has any recommendation. If not, the function returns the value zero. Otherwise, you can retrieve the recommendations using the [ ] operator. I added all the target prices to a temporary vector of prices and used the std::accumulate algorithm to compute the sum of all prices. Finally, the member function returns the total divided by the number of such price recommendations, which is just the average target price as desired.

Complete Code

Listing 4-1 presents the complete code for the class Recommendation, as described in the previous section. The listing shows the header and implementation files that you will need to include the class in your project.

Listing 4-1. Definitions and Implementation for Class Recommendation

```
//
//  Recommendation.h

#ifndef __FinancialSamples__Recommendation__
#define __FinancialSamples__Recommendation__

#include <string>

enum RecommendationType {
    BUY,
    SELL,
    HOLD,
    NO_RECOMMENDATION
};

class Recommendation {
public:
    Recommendation();
    Recommendation(const std::string &ticker, RecommendationType rec,
    double target);
    ~Recommendation();
    Recommendation(const Recommendation &r);
    Recommendation &operator =(const Recommendation &r);

    double getTarget() const;
    RecommendationType getRecommendation() const;
    const std::string &getTicker() const;

private:
    std::string m_ticker;
    RecommendationType m_recType;
    double m_target;
};

#endif /* defined(__FinancialSamples__Recommendation__) */

//
//  Recommendation.cpp
```

```cpp
#include "Recommendation.h"

Recommendation::Recommendation()
: m_recType(HOLD),
  m_target(0)
{

}

Recommendation::Recommendation(const std::string &ticker,
RecommendationType rec, double target)
: m_ticker(ticker),
  m_recType(rec),
  m_target(target)
{

}

Recommendation::~Recommendation()
{

}

Recommendation::Recommendation(const Recommendation &r)
: m_ticker(r.m_ticker),
  m_recType(r.m_recType),
  m_target(r.m_target)
{

}

Recommendation &Recommendation::operator =(const Recommendation &r)
{
    if (this != &r)
    {
        m_ticker = r.m_ticker;
        m_recType = r.m_recType;
        m_target = r.m_target;
    }
```

```
    return *this;
}

double Recommendation::getTarget() const
{
    return m_target;
}

RecommendationType Recommendation::getRecommendation() const
{
    return m_recType;
}

const std::string &Recommendation::getTicker() const
{
    return m_ticker;
}
//
//  RecommendationsProcessor.h

#ifndef __FinancialSamples__RecommendationsProcessor__
#define __FinancialSamples__RecommendationsProcessor__

#include <map>
#include <vector>

#include "Recommendation.h"

class RecommendationsProcessor {
public:
    RecommendationsProcessor();
    ~RecommendationsProcessor();
    RecommendationsProcessor(const RecommendationsProcessor &);
    RecommendationsProcessor &operator =(const RecommendationsProcessor &);

    void addRecommendation(const std::string &ticker, RecommendationType
    rec, double
    targetPrice);
    double averageTargetPrice(const std::string &ticker);
```

```
    RecommendationType averageRecommendation(const std::string &ticker);
private:
    std::map<std::string, std::vector<Recommendation> > m_recommendations;
};

#endif /* defined(__FinancialSamples__RecommendationsProcessor__) */

//
//  RecommendationsProcessor.cpp

#include "RecommendationsProcessor.h"
#include <numeric>

RecommendationsProcessor::RecommendationsProcessor()
{

}

RecommendationsProcessor::~RecommendationsProcessor()
{

}

RecommendationsProcessor::RecommendationsProcessor(const
RecommendationsProcessor &r)
: m_recommendations(r.m_recommendations)
{

}

RecommendationsProcessor &RecommendationsProcessor::operator
=(const RecommendationsProcessor &r)
{
    if (this != &r)
    {
        m_recommendations = r.m_recommendations;
    }
    return *this;
}
```

```cpp
void RecommendationsProcessor::addRecommendation(const std::string &ticker,
                                                 RecommendationType rec,
                                                 double targetPrice)
{
    Recommendation r(ticker, rec, targetPrice);
    m_recommendations[ticker].push_back(r);
}

double RecommendationsProcessor::averageTargetPrice(const std::string &ticker)
{
    if (m_recommendations.find(ticker) == m_recommendations.end())
        return 0;
    auto vrec = m_recommendations[ticker];
    std::vector<double> prices;
    for (auto i=0; i<vrec.size(); ++i)
    {
        prices.push_back(vrec[i].getTarget());
    }
    return std::accumulate(prices.begin(), prices.end(), 0) / prices.size();
}

RecommendationType RecommendationsProcessor::averageRecommendation(const
std::string &ticker)
{
    double avg = 0;
    if (m_recommendations.find(ticker) == m_recommendations.end())
    {
        return RecommendationType::NO_RECOMMENDATION;
    }
    auto vrec = m_recommendations[ticker];
    std::vector<int> recommendations;
    for (auto i=0; i<vrec.size(); ++i)
    {
        recommendations.push_back((int)vrec[i].getRecommendation()+1);
    }

    return (RecommendationType) (int) (avg / recommendations.size());
}
```

Performing Time-Series Transformations

Create a class that can be used to perform common time-series transformations, such as adding or subtracting values to prices and removing undesired values.

Solution

Time-series data filtering is the task of identifying and removing values, both for short- and long-term trends, in a sequence of data points. From the point of view of programming, this is performed by the application of a series of transformations to data stored in containers. This is a very common task that is properly covered by the STL. The STL provides several templates that simplify the execution of common algorithms such as sorting and selection, which can be applied to data containers such as vectors and lists. Such algorithms can be accessed in C++ code by including the header file <algorithm>. No additional work is necessary on the part of the programmer.

The <algorithm> header file provides declarations for many useful functions. Among them, you will find

- copy: This template function is used to copy a range of elements from a given container into a second container. Notice that, as with other generic algorithms, the containers don't need to be of the same type. For example, a range of a vector can be copied into a map, and vice versa.

- copy_backward: Similar to the copy function, but the process is performed from the last to first element of given range. This can be used, for example, to write the elements of a container in the reverse order.

- for_each: This algorithm can be used to apply a particular function or function object to a set of elements in a container. The for_each algorithm can be used to avoid the use of a for loop over the elements of a container. If the operation defined by the loop can be quickly encapsulated within a function or function object, the for_each algorithm can be a more concise way to perform the same operation.

- find_if: Used to find elements in a given range of a generic container. The last parameter for the find algorithm can be a function or a function object that is used to determine the property satisfied by the desired element.

125

- `count`: Return the number of elements in a given range of a generic container.

- `count_if`: Return the number of elements in a given range of generic containers that satisfy the function or function object passed as the third parameter.

- `transform`: This generic function takes a range of elements in a container, a destination container, and a transformation function object. The elements of the input container are transformed using the transformation function and stored in the destination container.

- `fill`: This function is used to fill a given range of a container with a single element.

- `reverse`: Changes the given range of a generic container so that the order of the elements in the container is reversed.

- `sort`: A generic algorithm that can be used to sort a sequence of elements stored in an STL container. The first two parameters for this algorithm define the range of elements. The third parameter is a comparison function, which is used to determine if two elements are correctly ordered.

- `binary_search`: Implements a binary search of elements over a sorted range in a container.

- `min_element`: Returns the element with minimum value among the elements stored in the given container.

- `max_element`: Returns the element with maximum value among the elements stored in the given container.

These are the most common algorithms provided by the STL. While these algorithms are in most cases simple to code, there are some advantages to using the STL algorithm templates over manually created implementations:

- The first advantage is that they reduce the possibility of errors when implementing similar operations. For example, the `for_each` algorithm just applies the same function to all elements in a range. While this is easy to do with a `for` loop, there is always the possibility of making mistakes when manipulating individual elements of the container. The `for_each` algorithm, however, has all the logic contained in the template definition, reducing the possibility of errors.

- Algorithms in the STL have intimate knowledge of how containers work. Implementers of the STL understand subtle nuances of the containers, which can greatly improve the performance of these algorithms. By using partial specialization, the authors of the STL can tailor such algorithms to achieve maximum performance for each container. Using the STL, you automatically take advantage of this knowledge in your application.

- Algorithms are also a succinct way to describe your intent. Instead of writing another `for` loop to find a minimum element, for example, you can apply the `min_element` algorithm to the target container.

Using STL Algorithms

To solve the problem posed in this section, you will create a class, `TimeSeriesTransformations`, which implements a few time-series transformation operations. The first algorithm implemented is used to reduce prices in the series. The solution relies on the `std::transform` algorithm, and it is implemented as follows:

```
void TimeSeriesTransformations::reducePrices(double val)
{
    std::vector<double> res;
    std::transform(m_prices.begin(), m_prices.end(), res.begin(),
        std::bind2nd(std::minus<double>(), val));
    m_prices.swap(res);
}
```

In this member function, the first step is to apply `std::transform` to the vector `m_prices`. The first two parameters are the iterators for the beginning and the end of the vector. Then, you need to pass the beginning of the output vector. Finally, the last parameter is the function object used to perform the transformation. The template `std::bind2nd` is declared in the `<functional>` header file and allows one to bind the second parameter of a functional object (in this case, the minus function). The result is that the minus function will be applied to the m_prices vector, with the second parameter set to the value defined by val. After the transformation is performed, the next step is to swap the values stored in m_prices with the values stored in the result vector.

The second function is similar, but I used an alternative strategy.

```
void TimeSeriesTransformations::increasePrices(double val)
{
    std::for_each(m_prices.begin(), m_prices.end(), std::bind1st(std::plus
    <double>(), val));
}
```

Here, the function template std::for_each is used to perform a transformation to each element of the original vector. In this way, you can avoid the need to swap values into and out of the container. The for_each function applies the plus function with the first parameter bound to the passed value. As a result, all prices are increased as desired.

The TimeSeriesTransformations class also includes a few other methods that explore the STL algorithms. For example, the removePricesLessThan method uses the remove_if template to eliminate prices that are less than the given value. The member function removePricesGreaterThan is similar. Finally, the getFirstPriceLessThan function uses the find_if template function to identify a price that is less than a given value, if such a price exists.

Complete Code

The algorithm described previously has been implemented using a C++ class called TimeSeriesTransformations. It is divided into a header file TimeSeriesTransformations.h and a source file TimeSeriesTransformations.cpp. Listing 4-2 contains the complete code.

Listing 4-2. Class TimeSeriesTransformations

```
//
//   TimeSeriesTransformations.h

#ifndef __FinancialSamples__TimeSeriesAnalysis__
#define __FinancialSamples__TimeSeriesAnalysis__

#include <vector>
```

```
class TimeSeriesTransformations {
public:
    TimeSeriesTransformations();
    TimeSeriesTransformations(const TimeSeriesTransformations &);
    ~TimeSeriesTransformations();
    TimeSeriesTransformations &operator=(const TimeSeriesTransformations &);
    void reducePrices(double val);
    void increasePrices(double val);
    void removePricesLessThan(double val);
    void removePricesGreaterThan(double val);
    double getFirstPriceLessThan(double val);
    void addValue(double val);
    void addValues(const std::vector<double> &val);
private:
    std::vector<double> m_prices;
};

#endif /* defined(__FinancialSamples__TimeSeriesAnalysis__) */

//
//   TimeSeriesTransformations.cpp

#include "TimeSeriesTransformations.h"

#include <algorithm>
#include <functional>

 TimeSeriesTransformations::TimeSeriesTransformations()
: m_prices()
{
}

TimeSeriesTransformations::TimeSeriesTransformations(const
TimeSeriesTransformations &s)
: m_prices(s.m_prices)
{
}
```

```
TimeSeriesTransformations::~TimeSeriesTransformations()
{
}

TimeSeriesTransformations &TimeSeriesTransformations::operator=(const
TimeSeriesTransformations &v)
{
    if (this != &v)
    {
        m_prices = v.m_prices;
    }
    return *this;
}

void TimeSeriesTransformations::reducePrices(double val)
{
    std::vector<double> neg(m_prices.size());
    std::transform(m_prices.begin(), m_prices.end(), neg.begin(),
        std::bind2nd(std::minus<double>(), val));
    m_prices.swap(neg);
}

void TimeSeriesTransformations::increasePrices(double val)
{
    std::for_each(m_prices.begin(), m_prices.end(), std::bind1st(std::plus
    <double>(), val));
}

void TimeSeriesTransformations::removePricesLessThan(double val)
{
    std::remove_if(m_prices.begin(), m_prices.end(), std::bind2nd(std::less
    <double>(), val));
}
```

```cpp
void TimeSeriesTransformations::removePricesGreaterThan(double val)
{
    std::remove_if(m_prices.begin(), m_prices.end(), std::bind2nd(std::
    greater<double>(), val));
}

double TimeSeriesTransformations::getFirstPriceLessThan(double val)
{
    auto res = std::find_if(m_prices.begin(), m_prices.end(),
        std::bind2nd(std::less<double>(), val));
    if (res != m_prices.end())
        return *res;
    return 0;
}

void TimeSeriesTransformations::addValue(double val)
{
    m_prices.push_back(val);
}

void TimeSeriesTransformations::addValues(const std::vector<double> &val)
{
    m_prices.insert(m_prices.end(), val.begin(), val.end());
}

int main()
{
    TimeSeriesTransformations ts;
    std::vector<double> vals = {7, 6.4, 2.16, 5, 3, 7};
    ts.addValues(vals);
    ts.addValue(6.5);
    ts.reducePrices(0.5);
    std::cout << " price is " <<  ts.getFirstPriceLessThan(6.0) << std::endl;
    return 0;
}
```

Running the Code

I included some sample code that uses the `TimeSeriesTransformations` class in the `main()` function. In this way, you can compile the files presented in the previous section into a sample application. After building the application with the help of a C++ compiler such as gcc or Visual Studio, you may run it using the following command line:

./TimeSeriesTransformations
```
 price is 5.9
```

Copying Transaction Files

Create a class to copy transaction files into a temporary storage.

Solution

File operations are a common requirement in a lot of areas of programming, and it couldn't be different in financial applications. Data is commonly stored in formats such as CSV and XML, as they need to be processed and filtered by other applications. Logging facilities are also necessary to guarantee that debugging and error messages are properly handled.

This type of file manipulation problem can be solved using traditional C interfaces, which are available for all major operating systems such as UNIX and Windows. However, such native interfaces have a few shortcomings and should be avoided when possible. I use this opportunity to provide an overview of a better approach, using the boost repository of C++ libraries, and the `filesystem` library in particular. With a basic knowledge of how boost works, you will be in a position to use other, more complex classes in the next few chapters.

Boost Libraries

The standard C++ library provides a large number of classes, containers, and algorithms. However, due to the substantial effort necessary to create a new standard of the C++ language, the included set of libraries is frequently minimalist, comprising only the essential functionality needed by most programmers. Moreover, the standard library incorporates only classes and functions that have been well verified and established by their use in real applications, having stood the test of time.

Due to the slow process of including new functionality in the language standard, a more agile strategy for software distribution and development was needed, as a way to better incorporate new libraries approved by the C++ community. Boost libraries were created to fill the gap left by this slow process of including new functionality in the standard. Unlike traditional single-vendor libraries, the goal for boost developers is to create high-level components that can be reusable across a large spectrum of domains and architectures. Many of the contributors to boost are themselves involved in the development of the C++ standard, which means that many of the libraries currently included in the boost distribution will later on become part of the standard library. For example, template classes such as `std::shared_ptr` originated from `boost::shared_ptr`.

To use boost libraries in your application, you need to download and install them from the `boost.org` website. The process is made simpler because of the nature of boost classes. Since most components in boost are defined as template classes, the complete code is, with a few exceptions, contained in the header files. This means that you can use all the functionality in certain boost libraries by simply adding a header file to your code.

Table 4-1 shows some of the libraries available in the boost distribution.

Table 4-1. *Some of the Most Commonly Used Components in the Boost Distribution*

Library	Description
boost:any	A polymorphic data type designed to be used as a container to any other data type
Bind	Allows existing functions to be used by other function objects
Circular Buffer	Defines a storage template that can be used as a circular buffer
Chrono	A set of time-related utilities
Filesystem	An implementation of common file operations, including coping, moving, and creating files or directories
Foreach	Introduces a looping construct (deprecated by new features on C++11)
Function	A set of templates that define function wrappers
Geometry	An implementation of common geometric algorithms

(*continued*)

Table 4-1. (*continued*)

Library	Description
Graph	Defines a graph data type, as well as common graph theory techniques and algorithms
Hash	A simple hash table data type, defined as a template
Lambda	A set of templates that can be used to write lambda functions for functional programming
Log	A generic logging facility for C++ applications
Math	Additional math functions
MPI	A message passing interface library
PropertyMap	Defines a generic property template, which can be used to define dynamic attributes
SmartPtr	A set of smart pointer types for storage of heap-allocated objects

In this chapter and the next, we will have the opportunity to explore some of these libraries, as they will be needed in other financial C++ code presented in this book. In this example, we are concerned with the filesystem library, used to provide file and directory-related operations.

C and C++ have traditionally provided interfaces for file handling in each of the platforms where it has been implemented. Platform vendors have created separate libraries for this purpose, resulting in different interfaces for operating systems such as UNIX and other Posix-compliant systems, Windows, OS/2, VMS, and others. The differences between these interfaces, however, make it difficult to port applications across systems. Application writers have, in practice, created abstraction layers that interact with each different system as needed.

The filesystem library in STL is an attempt to provide a set of cross-platform classes and templates for file manipulation. The same classes and templates can be used to handle files in each of the platforms supported by the STL. This reduces the amount of work by application programmers, while the resulting code can be reused in other platforms without risks.

The main components of the library are included in the std::filesystem namespace. These components allow you to perform common operations on files and directories.

The classes included in the library easily support operations such as copying, moving, changing permissions, and removing. Next, I show you some of these important components and how they can be used to write code to manipulate file system contents.

The `std::filesystem::path` class represents a path in the file system. Having a path class is interesting because it lets programmers represent directory paths in different systems while using the same object. For example, paths in a UNIX system use the "/" separator, while in Windows, the "\" separator is used. To avoid problems associated with these different conventions, the `filesystem` library uses a common representation. The class `path` is then used by many of the functions provided in the library.

The other important concept of the `filesystem` library is that of directory iterators. You can create an iterator for a particular path using the `directory_iterator` function. This function returns an iterator object, which has operators such as ++ and --, allowing programmers to move between elements of the given directory.

Other than suitable abstractions for paths and iterators, the `filesystem` library provides a set of functions that can be used to perform individual changes to files and directories. These functions include the following:

- `is_regular_file`: Returns true if the path supplied indicates a regular file (instead of a directory)

- `is_directory`: Returns true if the path indicates a directory

- `file_size`: Returns the size of the filename passed as argument, in bytes

- `exists`: Returns true if the path passed as argument exists in the file system

- `status`: Returns a `file_status` object, which encapsulates the properties of given file, such visibility and type

- `create_directory`: Creates a new directory in the file system, with the given path

- `copy`: Makes a copy of the given path to a location indicated by the second parameter

- `remove`: Removes the given path from the file system

- `current_path`: Returns a path object that indicates the current path used by the application

These functions can be easily combined to manipulate the file system. I used some of these functions to implement all the functionality needed by the FileManager class. For example, the following is how the getContents member function is coded:

```
std::vector<std::string> FileManager::getContents(const std::string
&prefix)
{
    std::vector<std::string> results;
    path aPath(prefix);
    if (!is_directory(aPath))
    {
        std::cout << " incorrect path was used " << std::endl;
    }
    else
    {
        std::vector <path> contents;
        copy(directory_iterator(aPath), directory_iterator(),
        back_inserter(contents));

        for (int i=0; i<contents.size(); ++i)
        {
            results.push_back(contents[i].string());
        }
    }
    return results;
}
```

The first step is to create a path object based on the string passed as a parameter. Once the path has been created, you can test if it points to a directory using the is_directory function. If the path is correct, then you can use the directory_iterator function to create an iterator object, which is passed to the copy function. The copy function's only job is to copy each element pointed by the iterator into the contents vector. The elements in this container are later converted to strings and added to the results vector.

Finally, the function that copies files from one directory to a given destination can be added to the `FileManager` class using the following code:

```
void FileManager::copyToTempDirectory(const std::string &prefix)
{
    path tmpPath("/tmp/");
    path aPath(prefix);
    if (!is_directory(aPath))
    {
        std::cout << " incorrect path was used " << std::endl;
        return;
    }
    std::cout << " copying the following files: " << std::endl;
    this->listContents(prefix);

    for (auto it = directory_iterator(aPath); it != directory_iterator();
    ++it)
    {
        if (is_regular_file(it->path()))
        {
            copy_file(it->path(), tmpPath);
        }
    }
}
```

Here, you start checking the given path prefix to make sure that it is a reference to a directory. Then, I have added some code to list the contents as a form of logging. The next step is to iterate through the content of the directory using the iterator returned by the `directory_iterator` function. For each element of the directory, you can test if it is a regular file and then use the `copy_file` function to perform the copy.

Complete Code

You can see the complete definition of the `FileManager` class in Listing 4-3. At the end of the listing, you can see an example of how to use the class in the `main()` function.

Listing 4-3. Definitions and Implementation for Class FileManager

```cpp
//
//  FileManager.h

#ifndef __FinancialSamples__FileManager__
#define __FinancialSamples__FileManager__

#include <string>
#include <vector>

class FileManager {
public:
    FileManager(const std::string &basePath);
    FileManager(const FileManager &);
    ~FileManager();
    FileManager &operator=(const FileManager &);

    void removeFiles();
    std::vector<std::string> getDirectoryContents();
    void listContents();
    void copyToTempDirectory(const std::string &prefix);

private:
    std::string m_basePath;
};

#endif /* defined(__FinancialSamples__FileManager__) */

//
//  FileManager.cpp

#include "FileManager.h"

#include <filesystem>
#include <iostream>

using namespace std::filesystem;
```

```cpp
FileManager::FileManager(const std::string &basePath)
: m_basePath(basePath)
{

}

FileManager::FileManager(const FileManager &v)
: m_basePath(v.m_basePath)
{

}

FileManager::~FileManager()
{

}

FileManager &FileManager::operator=(const FileManager &v)
{
    if (this != &v)
    {
        m_basePath = v.m_basePath;
    }
    return *this;
}

void FileManager::removeFiles()
{
    std::vector<std::string> files = getDirectoryContents();
    for (unsigned i=0; i<files.size(); ++i)
    {
        path aPath(files[i]);

        if (is_regular_file(aPath))
        {
            std::cout << " path " << files[i] << " is not a regular file "
            << std::endl;
        }
```

```
        else
        {
            remove(aPath);
        }
    }
}

void FileManager::listContents()
{
    std::vector<std::string> files = getDirectoryContents();
    for (unsigned i=0; i<files.size(); ++i)
    {
        path aPath(files[i]);
        if (is_regular_file(aPath))
        {
            std::cout << aPath.string() << std::endl;
        }
    }
}

std::vector<std::string> FileManager::getDirectoryContents()
{
    std::vector<std::string> results;
    path aPath(m_basePath);
    if (!is_directory(aPath))
    {
        std::cout << " incorrect path was used " << std::endl;
    }
    else
    {
        auto iterator = directory_iterator(aPath);

        std::vector <path> contents;
        copy(directory_iterator(aPath), directory_iterator(),
        back_inserter(contents));
```

```
        for (unsigned i=0; i<contents.size(); ++i)
        {
            results.push_back(contents[i].string());
        }
    }
    return results;
}
void FileManager::copyToTempDirectory(const std::string &tmpDir)
{
    const path tmpPath(tmpDir);
    path aPath(m_basePath);
    if (!is_directory(aPath))
    {
        std::cout << " incorrect path was used " << std::endl;
        return;
    }
    std::cout << " copying the following files: " << std::endl;
    this->listContents();

    std::vector<std::string> contents = getDirectoryContents();
    for (auto it = directory_iterator(aPath); it != directory_iterator();
    ++it)
    {
        if (is_regular_file(it->path()))
        {
            copy_file(it->path(), tmpPath);
        }
    }
}
int main()
{
    // create a FileManager object for the /tmp directory
    //
    FileManager fm("/tmp/");
    std::vector<std::string> contents = fm.getDirectoryContents();
    std::cout << "entries: " << std::endl;
```

```
    for (std::string entry : contents)
    {
        std::cout << entry << std::endl;
    }
    return 0;
}
```

Running the Code

The class `FileManager` presented in Listing 4-3 can be built using any standard C++ compiler, such as gcc or Visual Studio. No extra library will be needed to use the `std::filesystem` classes if you use a recent compiler (at least C++17). Many modern Linux distributions already include recent versions of `gcc`, but if you use other operating systems, check for support.

For example, the command line necessary to build this class using `gcc` in my system is

```
gcc -o FileManager FileManager.cpp
```

Once the application is generated, you can run it on a UNIX system as

```
./FileManager
```

This will display a list of files stored in the `/tmp` directory (you can change that directory as necessary to test on a path in your own system).

Handling Dates

Let's see how to create a class that can be used to determine trading days for common securities, which are negotiated from Monday to Friday.

Solution

Dates are such a common part of financial data that you should have a well-defined way to deal with them. Dates are an integral part of historical prices, as well as important events for equity analysis, such as earnings releases, dividends, price splits, and other regulatory actions. The same can be said about fixed income, derivatives, and other investment classes. C++ provides a wealth of features that can be used to store, calculate, and transform dates from one format to another.

Although there are many time- and date-related functions and classes in C++, many of these mechanisms have been inherited from C standard libraries and are not as easy to use as other components of the STL. To smooth this process of integration with the STL, the boost repository includes a `date_time` library that specializes in handling different representations of dates, as well as providing the basic support for calculations based on different date formats.

To solve the problem posed in this section, you will use a class called `Date`, which encapsulates the concept of date as used by the application. The member variables are simply three values representing the year, month, and day. There is also the concept of days of the week, which are encoded in an enumeration.

```
enum class DayOfWeek {
    Sun,
    Mon,
    Tue,
    Wed,
    Thu,
    Fri,
    Sat
};
```

The `Date` class exposes a number of member functions that can be used to answer common requests, such as `getDayOfWeek`, which returns the day of the week for the current date, and `isLeapYear`, which tells if a year has 29 days in February. Here is a quick list of member functions for `Date`.

```
Date(int year, int month, int day);
    ~Date();

    bool isLeapYear();
    Date &operator++();
    bool operator<(const Date &d);
    DayOfWeek getDayOfWeek();
    int daysInterval(const Date &);
    bool isTradingDay();
```

143

The `isLeapYear` method implementation uses the well-known definition of leap year, which considers years that are divisible by 4, 100, and 400:

```
bool Date::isLeapYear()
{
        if (m_year % 4 != 0) return false;
        if (m_year % 100 != 0) return true;
        if (m_year % 400 != 0) return false;
        return true;
}
```

The `getDayOfWeek` finds the day of the week for any date after January 1, 1900 (a Monday). It does so by counting the days since that date and updating the years, months, and days as necessary. The task of correctly adding to the current date is handled in `operator +`.

Finally, the `Date` class computes the difference between dates using the help of the `date_time` library from boost. In `date_time`, dates are classified according to a calendar. The calendar used in the Western world is called the Gregorian calendar. It can be used after you include the following header file:

```
<boost/date_time/gregorian/gregorian.hpp>
```

The `date_time` library defines a few generic data types that can be used for date manipulation. In this example, we are interested in the `date` and `date_duration` types. The `date` type is just a representation of a date and can be initialized with a year, month, and day. The `date_duration` is used to store the difference between dates. A duration type can be converted to an integer type using the `days()` member function. Here is how you can implement `daysInterval`:

```
int Date::daysInterval(const Date &d)
{
    date bdate1(m_year, m_month, m_day);
    date bdate2(d.m_year, d.m_month, d.m_day);

    boost::gregorian::date_duration duration = bdate1 - bdate2;
    return (int) duration.days();
}
```

Complete Code

You can see the complete code for the class Date in Listing 4-4. Listing 4-4 includes a header file and an implementation file for the class. You will also see a main() function that creates two objects of type Date and performs some simple operations with them.

Listing 4-4.   Implementation for Class Date

```
//
//  Date.h

#ifndef __FinancialSamples__Date__
#define __FinancialSamples__Date__

#include <string>

class Date {
public:

    enum class DayOfWeek {
        Sun,
        Mon,
        Tue,
        Wed,
        Thu,
        Fri,
        Sat
    };

    Date(int year, int month, int day);
    ~Date();

    bool isLeapYear();
    Date &operator++();
    bool operator<(const Date &d);
    DayOfWeek getDayOfWeek();
    int daysInterval(const Date &);
    bool isTradingDay();
    std::string toStringDate(Date::DayOfWeek day);
```

```cpp
private:
    int m_year;
    int m_month;
    int m_day;
};

#endif /* defined(__FinancialSamples__Date__) */

//
//  Date.cpp

#include "Date.h"

#include <vector>
#include <algorithm>

#include <boost/date_time/gregorian/gregorian.hpp>

using namespace boost::gregorian;

Date::Date(int year, int month, int day)
: m_year(year),
  m_month(month),
  m_day(day)
{
}

Date::~Date()
{
}

bool Date::isLeapYear()
{
        if (m_year % 4 != 0) return false;
        if (m_year % 100 != 0) return true;
        if (m_year % 400 != 0) return false;
        return true;
}
```

```cpp
Date &Date::operator++()
{
        std::vector<int> monthsWith31 = { 1, 3, 5, 7, 8, 10, 12 };

        if (m_day == 31)
        {
                m_day = 1;
                m_month++;
        }
        else if (m_day == 30 &&
                  std::find(monthsWith31.begin(),
                            monthsWith31.end(), m_month) == monthsWith31.end())
        {
                m_day = 1;
                m_month++;
        }
        else if (m_day == 29 && m_month == 2)
        {
                m_day = 1;
                m_month++;
        }
        else if (m_day == 28 && m_month == 2  && !isLeapYear())
        {
                m_day = 1;
                m_month++;
        }
        else
        {
                m_day++;
        }

        if (m_month > 12)
        {
                m_month = 1;
                m_year++;
        }
```

```
        return *this;
}

int Date::daysInterval(const Date &d)
{
    Date bdate1(m_year, m_month, m_day);
    Date bdate2(d.m_year, d.m_month, d.m_day);

    boost::gregorian::date_duration duration = bdate1 - bdate2;
    return (int) duration.days();
}

bool Date::operator<(const Date &d)
{
    if (m_year < d.m_year) return true;
    if (m_year == d.m_year && m_month < d.m_month) return true;
    if (m_year == d.m_year && m_month == d.m_month && m_day < d.m_day)
    return true;
    return false;
}

Date::DayOfWeek Date::getDayOfWeek()
{
    int day = 1;
    Date d(1900, 1, 1);
    for (;d < *this; ++d)
    {
        if (day == 7) day = 1;
        else day++;
    }
    return (DayOfWeek) day;
}

bool Date::isTradingDay()
{
    DayOfWeek dayOfWeek = getDayOfWeek();
```

```cpp
    if (dayOfWeek == DayOfWeek::Sun || dayOfWeek == DayOfWeek::Sat)
    {
        return false;
    }
    return true;
}

std::string Date::toStringDate(Date::DayOfWeek day)
{
    switch(day)
    {
        case DayOfWeek::Sun: return "Sunday";
        case DayOfWeek::Mon: return "Monday";
        case DayOfWeek::Tue: return "Tuesday";
        case DayOfWeek::Wed: return "Wednesday";
        case DayOfWeek::Thu: return "Thursday";
        case DayOfWeek::Fri: return "Friday";
        case DayOfWeek::Sat: return "Saturday";
    }
    throw std::runtime_error("unknown day of week");
}

int main()
{
    Date myDate(2015, 1, 3);
    auto dayOfWeek = myDate.getDayOfWeek();
    std::cout << " day of week is "
        << myDate.toStringDate(dayOfWeek) << std::endl;
    Date secondDate(2014, 12, 5);
    ++secondDate;  // test increment operator
    ++secondDate;

    int interval = myDate.daysInterval(secondDate);
    std::cout << " interval is " << interval << " days" << std::endl;
    return 0;
}
```

Running the Code

The code presented in Listing 4-4 uses standard classes that are available to any standard C++ compiler. It also uses the boost library, which is open source and can be downloaded for free from boost.org. You can build an application using the following command line on Linux and other UNIX systems (assuming that boost was installed on /usr/local/boost):

```
gcc -o Date -I/usr/local/boost/ Date.cpp
```

Executing the resulting binary will show the output of the test code included in the main() function:

```
./Date
 day of week is Saturday
 interval is 27 days
```

Conclusion

In this chapter, I presented a few programming examples that cover basic libraries used in financial programming. These include the STL, with its set of containers (such as vector, map) and algorithms (such as sort, transform, and for_each, among others). You have also learned about the boost repository, a group of libraries that has been created to fill the gap resulting from the slow standardization process in C++.

Algorithms are an extensive part of the STL. These algorithms can be used to perform common operations such as search, copy, and transform, in any container defined by STL templates. You have also learned in this chapter how to apply these algorithms to data containers in order to perform data analysis.

The first sample application shows how to handle analyst recommendations. To properly process this type of information, you had to use STL vectors and maps. The second sample application uses algorithms provided by the STL to perform simple transformations in a time series. This kind of transformation can be used to clean up data, perform what-if analysis, and update prices according to the requirements of new techniques for investment analysis. You have seen how this can be done using STL algorithms and functional templates.

You have also learned how to create C++ code that handles files and directories in a way that is independent of the platform or operating system. This is accomplished using the `filesystem` library, which is part of the boost repository. One of the main advantages of using boost is that, while other libraries are closely tied to the operating system, boost libraries are written in a platform-independent way, following the same strategies employed by the standard library. In fact, many components of the boost have become part of the standard library over the years.

Another important aspect of financial code is the frequent use of dates. This type of data is associated with trades, analyst recommendations, dividends, and so many other events related to an investment. You learned how to use the date type in the boost `date_time` library, as well as how to compute other interesting properties of dates.

This concludes a set of coding examples that reviews some basic aspects of modern C++ programming. You need to be aware of such techniques, which are mostly based on the use of templates, the STL, and their algorithms. It is also important to learn about extension libraries such as the boost repository. In the next chapter, you will start to learn more about the design of numerical classes. Financial applications in C++ make heavy use of numerical facilities to perform quick and accurate calculations of the desired properties of different investment classes. I will discuss some of the underlying principles in creating such numerical classes and how modern C++ libraries can help you simplify the resulting code.

CHAPTER 5

Designing Numerical Classes

At the heart of any high-performance financial application, there is a set of well-designed numerical classes. These classes are responsible for the implementation of concepts that are an integral part of tasks such as financial modeling, forecasting, and analysis of investment decisions. Without the support of mathematical models, it would be very difficult to propose and evaluate effective investment methodologies. That is why, as a programmer in the financial industry, you need to familiarize yourself with the best strategies to design and implement mathematically oriented code in C++. Although it is not necessary to become a math expert to use these programming techniques, it helps to possess a basic understanding of the most important numerical issues that need to be dealt with in your financial programming assignments.

This chapter will show you how to create classes that can run efficiently when used in numerically oriented, production-ready code. You will also see some sample code that show how to integrate existing numerical classes and algorithms into your applications.

Some of the concepts discussed in the code examples in this chapter include the following:

- How to design and implement an efficient matrix class

- How to supporting common matrix operations

- How to perform calculations at compilation time

- How to calculate factorial numbers using templates

- How to represent ratios as data types

- How to use and generate stochastic values using the boost library

153

© Carlos Oliveira 2021
C. Oliveira, *Practical C++20 Financial Programming*, https://doi.org/10.1007/978-1-4842-6834-6_5

Representing Matrices in C++

Implement a class that represents a matrix with some common associated operations, such as addition, subtraction, transposition, and multiplication.

Solution

Matrix manipulation is one of the basic operations in numerical computing. C++ doesn't have a matrix type, however, and it becomes necessary to implement matrices in most financial projects. The good news is that it is relatively easy to use algorithms already present in the standard templates library (STL) for this purpose, as you will see in the following coding example.

A matrix is just a two-dimensional arrangement of numbers, with which one can perform a set of standard mathematical transformations. In terms of data organization in memory, a matrix is not very different from a vector. Considering this similarity, we can take advantage of existing vector operations to facilitate the implementation of a Matrix class. Here is a possible definition for such a class.

```cpp
class Matrix {
public:
        typedef std::vector<double> Row;

    Matrix(int size, int size2);
    Matrix(int size);
    Matrix(const Matrix &s);
    ~Matrix();
    Matrix &operator=(const Matrix &s);

    void transpose();
    double trace();
    void add(const Matrix &s);
    void subtract(const Matrix &s);
    void multiply(const Matrix &s);

    Row & operator[](int pos);
private:
    std::vector<Row> m_rows;
};
```

Notice that, at the beginning of the public interface of the class, I used a public typedef to define a Row type. Since a row is just a vector of numbers, I want to avoid typing something as involved as std::vector<std::vector< > > to talk about simple rows. This is also a good measure that can help you avoid mistakes when defining new variables and member functions. As a result of this typedef, all the data stored in the matrix is declared as a vector of Row objects in the private section of the Matrix class.

The simple Matrix class just presented has two constructors.

```
Matrix(int size);
Matrix(int m, int n);
```

The first one creates a square matrix, that is, a matrix with the same number of rows and columns. The second constructor is used to create a more generic rectangular matrix, with m rows and n columns.

To make the matrix operate more like its counterpart, the vector, you can introduce a subscript operator to act as an access helper. In this way, it is possible to set and retrieve the value of specific entries in the matrix using native syntax. The implementation of the operator is straightforward, then, since we can refer to each individual row in the matrix.

```
Matrix::Row &Matrix::operator[](int pos)
{
    return m_rows[pos];
}
```

Next, we consider some elementary operations on matrices. The first operation is transposition, which is defined as the exchange of elements between rows and columns. That is, if A is a matrix, we need to interchange values between A[i][j] and A[j][i].

The second common operation on matrices is calculating the trace, which is defined as the sum of the elements in the main diagonal (i.e., those elements with the same row and column position). This can be implemented as follows:

```
double Matrix::trace()
{
    if (m_rows.size() != m_rows[0].size())
    {
        return 0;
    }
    double total = 0;
```

```
    for (unsigned i=0; i<m_rows.size(); ++i)
    {
        total += m_rows[i][i];
    }
    return total;
}
```

The first if statement checks if the matrix has a different number of rows and columns, in which case the trace operation is not defined. The for statement then iterates over the diagonal, adding those values to the total variable, which is returned at the end of the member function.

The Matrix class also implements the operations of adding and subtracting matrices. To add one matrix to another, you just need to add the individual elements of the first one to the corresponding elements in the second matrix. Similarly, the subtraction of matrices is defined element-wise. These operations are straightforward to implement in C++.

Finally, you can see how to implement matrix multiplication. In this case, you need to compute a new matrix, where each element is determined as the sum of the products of the i-th row and j-th column. The resulting matrix has dimensions determined by the number of rows in the current matrix and number of columns in the parameter matrix. The main part of the algorithm is the following:

```
std::vector<Row> rows;
for (unsigned i=0; i<m_rows.size(); ++i)
{
    std::vector<double> row;
    for (unsigned j=0; j<s.m_rows.size(); ++j)
    {
        double Mij = 0;
        for (unsigned k=0; k<m_rows[0].size(); ++k)
        {
            Mij += m_rows[i][k] * s.m_rows[k][j];
        }
        row.push_back(Mij);
    }
    rows.push_back(row);
}
m_rows.swap(rows);
```

156

In this code, we have three loops that range over the different dimensions of the original and the parameter matrix. The value `Mij` represents the element in position `[i][j]` for the resulting matrix. Notice that to simplify storage management, the algorithm performs the assignments in a new set of rows. Then, the results are stored in place of the existing values in the last line, using the swap function.

After the `Matrix` class has been defined, I have also added a few free operators that make it easier to work with the previously defined operations. These operators make sure that you can add, subtract, and multiply matrices using a syntax similar to that of native operations, although assuming a slight overhead for the temporary objects that become necessary. Here, for example, is the definition of `operator *`.

```
Matrix operator*(const Matrix &s1, const Matrix &s2)
{
        Matrix s(s1);
        s.multiply(s2);
        return s;
}
```

Complete Code

The ideas just described have been implemented in the `Matrix` class, presented here in Listing 5-1. This is a class that I will use in other examples in the next chapters of this book, so you should be familiar with its definition and main uses.

Listing 5-1. The Matrix Class

```
//
// Matrix.h

#ifndef __FinancialSamples__Matrix__
#define __FinancialSamples__Matrix__

#include <vector>

class Matrix {
public:
    typedef std::vector<double> Row;
```

```cpp
    Matrix(int size, int size2);
    Matrix(int size);
    Matrix(const Matrix &s);
    ~Matrix();
    Matrix &operator=(const Matrix &s);

    void transpose();
    double trace();
    void add(const Matrix &s);
    void subtract(const Matrix &s);
    void multiply(const Matrix &s);

    Row & operator[](int pos);
private:
    std::vector<Row> m_rows;
};

// free operators
//
Matrix operator+(const Matrix &s1, const Matrix &s2);
Matrix operator-(const Matrix &s1, const Matrix &s2);
Matrix operator*(const Matrix &s1, const Matrix &s2);

#endif /* defined(__FinancialSamples__Matrix__) */

//
//  Matrix.cpp

#include "Matrix.h"

Matrix::Matrix(int size)
{
    for (unsigned i=0; i<size; ++i )
    {
        std::vector<double> row(size, 0);
        m_rows.push_back(row);
    }
}
```

```cpp
Matrix::Matrix(int size, int size2)
{
    for (unsigned i=0; i<size; ++i )
    {
        std::vector<double> row(size2, 0);
        m_rows.push_back(row);
    }
}

Matrix::Matrix(const Matrix &s)
: m_rows(s.m_rows)
{
}

Matrix::~Matrix()
{
}

Matrix &Matrix::operator=(const Matrix &s)
{
    if (this != &s)
    {
        m_rows = s.m_rows;
    }
    return *this;
}

Matrix::Row &Matrix::operator[](int pos)
{
    return m_rows[pos];
}

void Matrix::transpose()
{
    std::vector<Row> rows;
    for (unsigned i=0;i <m_rows[0].size(); ++i)
    {
        std::vector<double> row;
```

```
        for (unsigned j=0; j<m_rows.size(); ++j)
        {
            row[j] = m_rows[j][i];
        }
        rows.push_back(row);
    }
    m_rows.swap(rows);
}

double Matrix::trace()
{
    if (m_rows.size() != m_rows[0].size())
    {
        return 0;
    }
    double total = 0;
    for (unsigned i=0; i<m_rows.size(); ++i)
    {
        total += m_rows[i][i];
    }
    return total;
}

void Matrix::add(const Matrix &s)
{
    if (m_rows.size() != s.m_rows.size() ||
        m_rows[0].size() != s.m_rows[0].size())
    {
        throw new std::runtime_error("invalid matrix dimensions");
    }
    for (unsigned i=0; i<m_rows.size(); ++i)
    {
        for (unsigned j=0; j<m_rows[0].size(); ++j)
        {
            m_rows[i][j] += s.m_rows[i][j];
        }
```

```
    }
}

void Matrix::subtract(const Matrix &s)
{
    if (m_rows.size() != s.m_rows.size() ||
        m_rows[0].size() != s.m_rows[0].size())
    {
        throw new std::runtime_error("invalid matrix dimensions");
    }
    for (unsigned i=0; i<m_rows.size(); ++i)
    {
        for (unsigned j=0; j<m_rows[0].size(); ++j)
        {
            m_rows[i][j] += s.m_rows[i][j];
        }
    }
}

void Matrix::multiply(const Matrix &s)
{
    if (m_rows[0].size() != s.m_rows.size())
    {
        throw new std::runtime_error("invalid matrix dimensions");
    }
    std::vector<Row> rows;
    for (unsigned i=0; i<m_rows.size(); ++i)
    {
        std::vector<double> row;
        for (unsigned j=0; j<s.m_rows.size(); ++j)
        {
            double Mij = 0;
            for (unsigned k=0; k<m_rows[0].size(); ++k)
            {
                Mij += m_rows[i][k] * s.m_rows[k][j];
            }
```

```
                row.push_back(Mij);
        }
        rows.push_back(row);
    }
    m_rows.swap(rows);
}

Matrix operator+(const Matrix &s1, const Matrix &s2)
{
        Matrix s(s1);
        s.subtract(s2);
        return s;
}

Matrix operator-(const Matrix &s1, const Matrix &s2)
{
        Matrix s(s1);
        s.subtract(s2);
        return s;
}

Matrix operator*(const Matrix &s1, const Matrix &s2)
{
        Matrix s(s1);
        s.multiply(s2);
        return s;
}
```

Using Templates to Calculate Factorials

In this section, I will show how to create a template-based class that can be used to calculate factorials at compile time.

Solution

Templates provide an easy way to apply the same code across data types, allowing programmers to create generic, reusable code. The best example of this is the STL, with its many containers and associated algorithms. However, templates can also be used to perform numerical tasks due to their ability to receive integer numbers, in addition to data types, as formal arguments. In this coding example, you will see how templates can be employed to perform some simple calculations at compilation type.

Template-based computation can be seen as a useful strategy to reduce the runtime overhead of numeric algorithms. After all, if you're able to perform some of the calculations at compilation time, less time will be necessary to perform the complete computation each time you execute the compiled code.

One of the biggest surprises for people who start working with template-based computing is that calculated values cannot simply be returned as the output of functions. Since functions can return any value at runtime, a traditional function cannot serve as the basis for compile-time calculations. Instead, you need a way to store values inside the class as a constant, which can then become readily available to the compiler. One of the ways to achieve this in C++ is with an enumeration. For example, consider

```
enum {
   result = 1
};
```

This fragment defines a constant, integral value that can be later referenced in the program. If a constant expression is used (instead of a number) in the right-hand side of the declaration, the `result` value can be later employed in the program to access the desired value.

The next thing you need is a way to pass numbers as parameters to the class template. In C++, you can declare templates that take as parameters an `int` value (or one of its several variations such as `long` and `char`). The general syntax that can be used to perform calculations as part of a template is the following:

```
template <int N>
class CompileTime {
public:
   enum {  result = ConstantExpressionDependingOnN };
};
```

where `ConstantExpressionDependingOnN` is an expression that in some way depends on the parameter N and can be used to calculate the desired value. You can see that the code in this example will use this general format to perform compile-time calculations.

Once you find a way to execute calculations at compilation time, the next step is to introduce concepts such as iteration to your code. In C++ templates, it is not possible to write loops, such as `for` or `while`, as part of a constant expression. All C++ loops are executed at runtime, which makes them unusable for compile-time operations. Thankfully, templates provide a specialization mechanism that can be used to implement recursion, a technique that can be used to achieve the same effects as looping.

For example, if a template uses a single integer parameter, you can specialize that parameter with a base case alongside a generic version that handles the common case. Together, these cases are enough to simulate a loop that starts with the generic case and terminates the computation once the special case is reached. Figure 5-1 presents an illustration of this mechanism, where the following example is considered:

```
// general case
template <int N>
class Double {
public:
  enum {  result = 2 + Double<N-1>  };
};

// specialization for the base case
template <>
class Double<1> {
public:
  enum {  result = 2 };
};
```

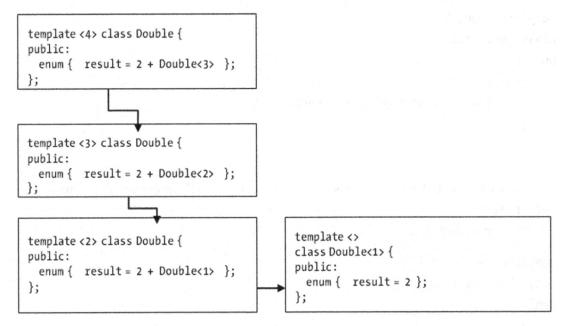

```
template <4> class Double {
public:
  enum {  result = 2 + Double<3>  };
};
```

```
template <3> class Double {
public:
  enum {  result = 2 + Double<2>  };
};
```

```
template <2> class Double {
public:
  enum {  result = 2 + Double<1>  };
};
```

```
template <>
class Double<1> {
public:
  enum {  result = 2 };
};
```

Figure 5-1. *An example of computation using template specialization. The general case is instantiated with the integer 4, and new instantiations are used until the specialization for Double<1> is reached*

This shows how you can compute the double of an integer number using template-based recursion. The general case is stated at the top, where the result value is defined as the expression 2 + Double<N-1>.

To find the value of that expression, the compiler will need to expand it inline, decreasing the value of N each time and calling Double with the new value. The second part of the declaration allows this process to end, introducing a base value. The declaration reads

```
template <>
class Double<1>
```

This tells the compiler that Double<1> is a specialization of a generic template, for the particular value of 1. Therefore, when the template Double is applied to 1, the result calculated will be the value 2, as desired.

A similar strategy can be used to solve multiple problems, including the required task of computing factorials. The first part is to define the general case, which contains a recurring expression.

```
template <long N>
class Factorial {
public:
    enum {
        result = Factorial<N-1>::result * N
    };

};
```

The second part of the solution is the base case, which will determine the value of Factorial<0>.

This can be written as

```
template <>
class Factorial<0> {
public:
    enum {
        result = 1
    };
};
```

We can test the preceding code with a few calls to the Factorial template. The example is included as part of the showFactorial function.

```
void showFactorial()
{
    std::cout << " Some factorial values: " << std::endl;
    std::cout << "fact(5)= " << Factorial<5>::result;
    std::cout << "fact(7)= " << Factorial<7>::result;
    std::cout << "fact(9)= " << Factorial<9>::result;
}
```

Finally, you can also use the Factorial class as the basis of other compile-time computations. For example, here is how you can use Factorial to calculate the choice number (the number of combinations of N objects, taken in groups of P).

```
template <int N, int P>
class ChoiceNumber {
public:
    enum {
        result = Factorial<N>::result / (Factorial<P>::result *
        Factorial<N-P>::result)
    };
};
```

Notice that here you don't need a base case, since there is actually no recursion involved. The ChoiceNumber class template is just using Factorial directly to perform the compile-time calculation as its result.

The nice thing about compile-time computations using templates is that you can use the same strategy discussed here to compute very different functions. As long as you can represent the computation as a recursion, the scheme described previously can be employed with little modification. In this way, you will be using the power of the compiler to perform calculations ahead of time and possibly saving a lot of time later, when the program is actually running.

Caution   While the ability of calculating values using templates is very useful, you may want to avoid using them frequently, since they may slow down compilation. Long compilation times may be the biggest adverse effect of overreliance on templates. Ideally, you should consider the trade-off between compilation time and runtime savings before deciding if a computation should be performed using templates at compilation time instead of runtime.

Complete Code

Here you have an example of using templates to calculate factorials at compile time. The important part of the implementation is in the header file (FactorialTemplate.h) shown in Listing 5-2. This is necessary, since templates need to be visible to the client at the moment they are used. The cpp file shows some sample uses of the template.

Listing 5-2. The FactorialTemplate.h Header File

```
//
//  FactorialTemplate.h

#ifndef __FinancialSamples__FactorialTemplate__
#define __FinancialSamples__FactorialTemplate__

template <long N>
class Factorial {
public:
    enum {
        result = Factorial<N-1>::result * N
    };
private:

};

template <>
class Factorial<0> {
public:
    enum {
        result = 1
    };
};

template <int N, int P>
class ChoiceNumber {
public:
    enum {
        result = Factorial<N>::result / (Factorial<P>::result *
        Factorial<N-P>::result)
    };
};

void showFactorial();

#endif /* defined(__FinancialSamples__FactorialTemplate__) */
```

```
//
// FactorialTemplate.cpp

#include "FactorialTemplate.h"

#include <iostream>

void showFactorial()
{
    std::cout << " Some factorial values: " << std::endl;
    std::cout << "fact(5)= " << Factorial<5>::result;
    std::cout << "fact(7)= " << Factorial<7>::result;
    std::cout << "fact(9)= " << Factorial<9>::result;
}

int main(int argc, const char **argv)
{
    std::cout << "factorial(6) = " << Factorial<6>::result;
    std::cout << "\n choiceNumber(5,6) = " <<  <<
    ChoiceNumber<6,2>::result;
    showFactorial();
    return 0;
}
```

Running the Code

You can compile and run the FactorialTemplate class to test the concepts you have just learned. For that purpose, you can use the gcc compiler, which can generate an application using the following command:

gcc -o factorial FactorialTemplate.cpp

After a few seconds, the binary file factorial will be created with the desired functionality. You can run the program by just calling it from the command line

./factorial

You can also click the executable if running on a Windows machine. This would result in the following output:

```
factorial(6)= 720
choiceNumber(5,6) = 15
factorial(5)= 120
factorial(7)= 5040
factorial(9)= 362880
```

Using C++20 Features to Compute Factorial

C++ has recently introduced a number of features that make it easier to work with template code. While the previous example is still useful to explain how templates work, the syntax has been simplified, and it is now possible to achieve the same results with much less boilerplate.

First, C++ now has the ability to calculate expressions at compilation time using the constexpr keyword. When using the constexpr keyword, you're instructing the compiler to directly perform a calculation during compilation time. This is much simpler than creating a recursive template class as you have seen in the previous section.

So, for example, if you need to create a factorial function, the following definition would be enough:

```
constexpr int factorial(const int n)
{
    return n <= 1 ? 1 : (n * factorial(n - 1));
}
```

The definition of this function uses constexpr, which means that its value should be calculated at compilation time, if possible (i.e., if the arguments are constant values known to the compiler). The remaining of the syntax is similar to what you would write for a standard function, using a recursive call to compute the value of factorial, based on the result for a smaller integer number. You can also call this function in the same way that you call a normal function.

Representing Calmar Ratios at Compile Time

The Calmar ratio is a measure of investment returns as compared to possible annual losses. It is used to compare investments with different risk profiles. The Calmar ratio is defined as the average annual rate of return for a given period, divided by the maximum drawdown (i.e., the maximum loss) during the same period. If you consider the same rate of return, investments with higher Calmer ratio had lower risk during the considered period. In this section, I will show how to create C++ code to represent Calmar ratios using compile-time techniques.

Solution

When writing numerical algorithms, it is frequently useful to represent certain quantities as constants. Some of these mathematical constants, however, are better denoted as ratios. For example, physical quantities frequently employ units of measurement, which are regularly represented as the ratio of other more fundamental units. As a consequence, ratios are a specific type of mathematical constant that can benefit from a more specific, high-level representation.

In this coding example, you will solve this problem using a simple library that is part of the boost repository. The library is called `ratio` and uses templates to represent mathematical quantities such as the standard Calmar ratio of an investment. The used representation can also be checked during compilation.

The basic template provided in the `ratio` library is simply called `ratio`. Its operation requires two template parameters, respectively, the numerator and the denominator. These parameters can either be simple numeric types, such as `int`, or other types previously declared using the `ratio` template. Types defined with the `ratio` template for different inputs are fundamentally different, and the compiler will enforce the correctness of any arithmetic operations involving these values.

One of the main advantages of using the `ratio` template is that it also provides some common compile-time operations. These operations can be used to perform standard mathematical transformations to the quantities defined with the template. Such operations include

- boost::ratio_add

- boost::ratio_subtract

- boost::ratio_multiply

- boost::ratio_add

- boost::ratio_negate

Using these operations, you can define derived types and constants, which are derivatives of the original ratio types. You can also use a few template-based operations to perform logical comparisons on such ratios, such as

- boost::ratio_equal
- boost::ratio_not_equal
- boost::ratio_less
- boost::ratio_less_equal
- boost::ratio_greater
- boost::ratio_greater_equal

You can start using the ratio library by importing the main header file `<boost/ratio.hpp>`. Then, you can start defining objects for each desired ratio using `boost::ratio`.

```
#include <boost/ratio.hpp>

boost::ratio<1, 2> one_half;
boost::ratio<1, 3> one_third;
boost::ratio<2, 5> two_fifths;
```

Once a `boost::ratio` object has been defined, you can retrieve its information at runtime using the `num` and `den` member variables, which correspond to the numerator and denominator, respectively. For example:

```
std::cout << "one_third numerator: " << one_third.num
          << " denominator: " << one_third.den;
```

Representing Calmar Ratios

With the help of the ratio library, it is possible to create a few useful financial types such as a Calmar ratio. The Calmar ratio is defined as the annual rate of return of an investment divided by its maximum drawdown during the known period. Thus, a `CalmarRatio` type can be defined as follows:

```
typedef boost::ratio<1, 1>::type CalmarRatioType;
```

From now on, `CalmarRatioType` can be used to represent quantities with a numerator and denominator at compile time. More interestingly, suppose that we want to be able to represent Calmar ratios using percentages as well as percentage points (1/100%). The definitions would then become

```
typedef boost::ratio<1, 100>::type CalmarRatioBPS;
typedef boost::ratio<1, 1>::type CalmarRatioPerc;
```

With these two types, we could create a template-based class to return information about the particular ratio, such as the maximum drawdown and the performance of the given object. The implementation is as follows:

```
template <class Ratio>
class CalmarRatio {
public:
        CalmarRatio(double calmar, double ret) : m_calmar(calmar),
        m_return(ret) {}
        virtual ~CalmarRatio() {}

        double getReturn();
        double getDrawDown()
        {
                return m_return / m_calmar * m_ratio.den;
        }
private:
        Ratio m_ratio;
        double m_calmar;
        double m_return;
};
```

The class is a template that receives the desired ratio type, either `CalmarRatioPerc` or `CalmarRatioBPS`. Of course, other ratio types could be supported if needed. Let's check the `getDrawDown` member function. The standard definition uses the den variable to calculate the drawdown of the investment. However, different versions of this member function can be created using template specializations. The following implementation provides an example:

```
template <>
double CalmarRatio<CalmarRatioBPS>::getDrawDown()
{
        return  m_return / m_calmar * m_ratio.den * 100;
}
```

In this case, since the template is specialized for the CalmarRatioBPS, the standard drawdown is multiplied by 100. This is necessary because the denominator is expressed in basis points, instead of percentages.

Complete Code

Listing 5-3 presents an implementation for the CalmarRatio class. Notice the use of the boost::ratio template to model different ratio types and how they are used by the main class.

Listing 5-3. The CalmarRatio Class

```
// CalmarRatio.h
//

#ifndef CALMARRATIO_H_
#define CALMARRATIO_H_

#include <boost/ratio.hpp>

typedef boost::ratio<1, 1>::type CalmarRatioType;
typedef boost::ratio<1, 100>::type CalmarRatioBPS;
typedef boost::ratio<1, 1>::type CalmarRatioPerc;

template <class Ratio>
class CalmarRatio {
public:
        CalmarRatio(double calmar, double ret) : m_calmar(calmar),
        m_return(ret) {}
        ~CalmarRatio() {}
```

```cpp
        double getReturn()
        {
                return m_return;
        }

        double getDrawDown()
        {
                return m_return / m_calmar * m_ratio.den;
        }
private:
        Ratio m_ratio;
        double m_calmar;
        double m_return;
};

template <>
double CalmarRatio<CalmarRatioBPS>::getDrawDown()
{
        return  m_return / m_calmar * m_ratio.den * 100;
}

#endif /* CALMARRATIO_H_ */

// CalmarRatio.cpp
//

#include "CalmarRatio.h"

#include <iostream>

boost::ratio<1, 2> one_half;
boost::ratio<1, 3> one_third;

void createCalmarRatio()
{
        CalmarRatio<CalmarRatioPerc> ratio(0.15, 11);
}
```

```cpp
void printRatios()
{
        std::cout << "one_third numerator: " << one_third.num
                << " denominator: " << one_third.den;
}

int main()
{
        CalmarRatio<CalmarRatioPerc> ratio(0.110, 3.12);
        std::cout << "return: " << ratio.getReturn()
                << " drawdown: " << ratio.getDrawDown() << std::endl;

        CalmarRatio<CalmarRatioBPS> bpsRatio(480, 2.15);
        std::cout << "return: " << bpsRatio.getReturn()
                << " drawdown: " << bpsRatio.getDrawDown() << std::endl;

}
```

Running the Code

We tested the sample application containing the `CalmarRatio` class and its associated code in a UNIX system using the gcc compiler. You can compile the cpp file presented earlier using a build system such as make, with its related makefile. Or you can just build the application directly with the compiler, using the following command line:

```
gcc -o calmarRatio CalmarRatio.cpp
```

The resulting executable file can be called from the terminal. It will display the result of the Calmar ratios in the following way:

```
return: 3.12 drawdown: 28.3636
return: 2.15 drawdown: 44.7917
```

As you see, the code treats the parameters differently, with results interpreted according to the type of Calmar ratio used. The first example uses a `CalmarRatioPerc`, which regards the Calmar ratio as applied to a percentage. The second example uses a `CalmarRatioBPS` representation, which works with basis points instead of percentages. The results, however, are displayed correctly according to their respective return and drawdown.

Generating Statistical Data

Create data using statistical distributions such as Gaussian (normal distribution) and chi-squared.

Solution

When working with trading algorithms, it is frequently useful to test the operation of such strategies on artificially generated prices. If we consider that most of the short-term movements of the market have a stochastic component, we can use random number generators to approximate the values of typical price-related time series.

In this section, we investigate how to generate statistically based data that can later be used to test trading strategies. To do this, you can use one of the many libraries currently available for the generation of statistical values in C++. These libraries operate similarly to random number generators, with a few differences. Traditional random number generators are used to produce random integer values. Such numbers can be, with some work, converted into uniformly distributed random numbers in a given interval, such as between 0 and 1.

For more advanced uses, however, it is interesting to generate random numbers from a particular probability distribution. Such probability distributions are based on standard random processes and include the Gaussian distribution (also known as normal distribution) and the chi-squared distribution (a form of skewed normal distribution). See Figure 5-2. These distributions can be used to generate stochastic numbers that are more representative of the stock market.

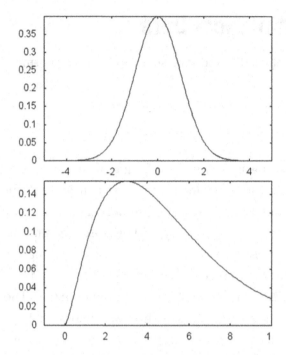

Figure 5-2. *Plots for two probability distribution functions (PDF) available in the* `boost::math` *namespace. The top plot is for a normal distribution with parameters 0 and 1. The bottom plot is for the chi-squared distribution with parameter 5*

In this example, I will show you how to use the `boost::math` namespace, where you declare a set of objects representing statistical distributions. The use of a library for this task makes it possible to concentrate on the design of your algorithm, instead of having to re-implement such a common statistic utility function, which has already been made available in many programming libraries. I also use the `boost::random` library to generate random data points based on these distributions.

Probability Distributions

Before I start, let me give you a quick overview of probability distributions and their uses. Probability distributions are a mathematical representation of parameterized random processes that frequently occur in nature. For example, the most basic probability distribution is the **uniform** distribution, in which points occur with the same probability over the whole interval in which the function is defined. Thus, each time a new event occurs according to this distribution (assuming that it has been defined for numbers between 1 and 2), its value may be any real number between 1 and 2 with equal probability. The uniform distribution has an important role because values generated under uniform random probability can be converted into other probability distributions.

Another important distribution is normal (or Gaussian) distribution. The normal distribution has two parameters: the mean (average) value and the standard deviation. Normally distributed random events occur with highest probability around the mean, and the probability of an event occurring further away decreases quickly. The resulting probability distribution is bell-shaped to indicate this characteristic of the probability space. It has been observed that many natural phenomena follow the normal distribution, especially when large numbers of observations are considered. Figure 5-2(top) shows a plot of the probability distribution function (PDF) for a normal (also known as Gaussian) random variable.

Other probability distributions are also used in financial applications. You can see a quick list of the most important in Table 5-1. Each distribution has a common usage pattern and associated parameters that can be used to describe the range of probabilities as well as the shape of the resulting function.

Table 5-1. *A Few Commonly Used Distributions with Their Parameters and Corresponding boost::math Identifier*

Distribution	Parameter(s)	boost::math identifier
Bernoulli	Probability of success	`boost::math::bernoulli_distribution`
Beta	Alpha and beta (real values)	`boost::math::beta_distribution`
Binomial	Number of trials and success probability	`boost::math:: binomial_distribution`
Cauchy	Location and scale	`boost::math::cauchy_distribution`
Chi-squared	Degrees of freedom	`boost::math::chi_squared_distribution`
Exponential	Lambda (rate)	`boost::math::exponential_distribution`
Geometric	Success probability	`boost::math::geometric_distribution`
Hypergeometric	N, K, and number of trials	`boost::math::hypergeometric_distribution`
Log-normal	Mean and sigma	`boost::math::lognormal_distribution`
Logistic	Mean and scale	`boost::math::logistic_distribution`

(*continued*)

Table 5-1. (*continued*)

Distribution	Parameter(s)	boost::math identifier
Normal (Gaussian)	Mean and sigma	`boost::math::normal_distribution`
Poisson	Lambda (rate)	`boost::math::poisson_distribution`
Student's t-Distribution	Degrees of freedom (real value)	`boost::math::students_t_distribution`
Triangular	Extremes and middle point	`boost::math::triangular_distribution`
Uniform	Start and end of interval	`boost::math::uniform_distribution`

Some of these functions are also available in the STL under the header file `<random>`. However, for completeness, I also show in this chapter how to compute these values using the boost library. In order to use some of these probability distributions in your code, you can include the header file `<boost/math/distributions.hpp>`. First, you need to make sure that boost is properly installed in your system (check the installation instructions on the `www.boost.org` website). The last column of Table 5-1 lists the distribution names.

Once you import a particular distribution, you can use it to respond to common questions such as the following: What is the mean of the distribution? What is the respective quantile for a particular value? What is the CDF of a particular value? You will see some of these questions being answered in class `DistributionData`, which is listed here.

Another responsibility of class `DistributionData` is to generate random numbers for some distributions, given the required parameters. A distribution-specific random number is created when the distribution object is called. You need to pass a uniform random number generator, which is also provided by boost. You can store these values in a vector and return them at the end of the member function. Here is an example of how this process works for Gaussian-distributed data.

```
std::vector<double> DistributionData::gaussianData(int nPoints, double
mean, double sigma)
{
        std::vector<double> data;

        boost::random::normal_distribution<> distrib(mean, sigma);
```

```
for (int i=0; i<nPoints; ++i)
{
        double val = distrib(random_generator);
        data.push_back(val);
}

return data;
}
```

Two other common probability distributions are the gamma and log-normal distributions. The gamma distribution can be interpreted as a generalized version of the normal distribution, in which you can control the shape and scale of the probabilities. Figure 5-3 (top) shows an example of the gamma distribution. The log-normal distribution is another possible generalization of the normal distribution, and it can be interpreted as the product of several positive and independent random variables. Its PDF is presented in Figure 5-3 (bottom). The log-normal distribution is included as one of the distributions supported by class `DistributionData`.

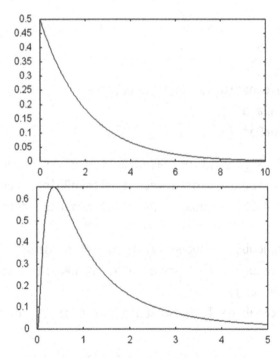

Figure 5-3. *Plots for two probability distribution functions (PDF) available in the boost::math namespace. The top plot is for a gamma distribution with parameters 1 and 2. The bottom PDF is for the log-normal distribution with parameters 0 and 1*

Complete Code

In Listing 5-4, I show the implementation for a class that generates data using a few of the probability distributions available in the boost::random template library. The main class is called DistributionData, and you can use it to generate numbers, as well as calculate quantiles for some distributions.

Listing 5-4. The DistributionData Class

```
// DistributionData.h
//

#ifndef DISTRIBUTIONDATA_H_
#define DISTRIBUTIONDATA_H_

#include <vector>

// class responsible for generating data basic on common probability
distributions
//
class DistributionData {
public:
        // standard constructor and destructor
        DistributionData();
        ~DistributionData();

        // random data generation based on the given parameters.
        // each function returns a vector with nPoints random values.
        std::vector<double> gaussianData(int nPoints, double mean, double
        sigma);
        std::vector<double> exponentialData(int nPoints, double rate);
        std::vector<double> chiSquaredData(int nPoints, int
        degreesOfFreedom);
        std::vector<double> logNormalData(int nPoints, double mean, double
        sigma);

        // returns the quantile of the give value x, corresponding to the
        parameters provided.
        //
```

```cpp
        double gaussianQuantile(double x, double mean, double sigma);
        double chiSquaredQuantile(double x, int degreesOfFreedom);
        double exponentialQuantile(double x, double rate);
        double logNormalQuantile(double x, double mean, double sigma);
};

#endif /* DISTRIBUTIONDATA_H_ */

// DistributionData.cpp
//

#include "DistributionData.h"

#include <boost/math/distributions.hpp>

using boost::math::quantile;

#include <boost/random.hpp>
#include <boost/random/normal_distribution.hpp>

static boost::rand48 random_generator;

DistributionData::DistributionData()
{
}

DistributionData::~DistributionData()
{
}

std::vector<double> DistributionData::gaussianData(int nPoints, double
mean, double sigma)
{
        std::vector<double> data;

        boost::random::normal_distribution<> distrib(mean, sigma);

        for (int i=0; i<nPoints; ++i)
        {
                double val = distrib(random_generator);
                data.push_back(val);
        }
```

```cpp
        return data;
}

std::vector<double> DistributionData::exponentialData(int nPoints, double
rate)
{
    std::vector<double> data;

        boost::random::exponential_distribution<> distrib(rate);

        for (int i=0; i<nPoints; ++i)
        {
                double val = distrib(random_generator);
                data.push_back(val);
        }

        return data;
}

std::vector<double> DistributionData::logNormalData(int nPoints, double
mean, double sigma)
{
    std::vector<double> data;

        boost::random::lognormal_distribution<> distrib(mean, sigma);

        for (int i=0; i<nPoints; ++i)
        {
                double val = distrib(random_generator);
                data.push_back(val);
        }

        return data;
}

std::vector<double> DistributionData::chiSquaredData(int nPoints, int
degreesOfFreedom)
{
        std::vector<double> data;
```

```
        boost::random::chi_squared_distribution<> distrib(degreesOfFreedom);

        for (int i=0; i<nPoints; ++i)
        {
                double val = distrib(random_generator);
                data.push_back(val);
        }

        return data;
}

double DistributionData::gaussianQuantile(double x, double mean, double
sigma)
{
        boost::math::normal_distribution<> dist(mean, sigma);

        return quantile(dist, x);
}

double DistributionData::chiSquaredQuantile(double x, int degreesOfFreedom)
{
        boost::math::chi_squared_distribution<> dist(degreesOfFreedom);

        return quantile(dist, x);
}

double DistributionData::exponentialQuantile(double x, double rate)
{
    boost::math::exponential_distribution<> dist(rate);

        return quantile(dist, x);
}

double DistributionData::logNormalQuantile(double x, double mean, double
sigma)
{
    boost::math::lognormal_distribution<> dist(mean, sigma);

        return quantile(dist, x);
}
```

```cpp
namespace {
    template <class T>
    void printData(const string &label, const T &data)
    {
        cout << " " << label << ":  ";
        for (auto i : data)
        {
            cout << i << " ";
        }
        cout << endl;
    }
}

int main()
{
    DistributionData dData;
    auto gdata = dData.gaussianData(10, 5, 2);
    printData("gaussian data", gdata);

    auto edata = dData.exponentialData(10, 4);
    printData("exponential data", edata);

    auto kdata = dData.chiSquaredData(10, 5);
    printData("chi squared data", kdata);

    auto ldata = dData.logNormalData(10, 8, 2);
    printData("log normal data", ldata);
    return 0;
}
```

Running the Code

You can compile the code in Listing 5-4 using any standard-compliant C++ compiler. You need to have boost installed in your system, as discussed in the previous sections. The following is an example of the expected output (exact numbers will vary depending on your particular implementation and random numbers used):

```
./distributionData
gaussian data:   7.12699 5.56941 5.91951 3.44111 4.89098 4.95243 7.33077
10.6359 5.00597 3.08975
exponential data:   0.108161 0.212945 0.0355506 0.0165794 0.753239 0.041679
0.219658 0.0610242 0.410622 0.0378433
chi squared data:   6.12073 2.14098 1.57523 6.49539 3.15154 1.47554 8.39545
9.07183 2.77768 5.05356
log-normal data:   1573.09 473.919 370.7 1212.54 1530.16 323705 2586.73
35919.6 628.913 372.41
```

Conclusion

Numerical classes and functions play a very important role in the development of financial engineering models. They offer the basic level of mathematical support needed for the creation of sophisticated trading strategies. In this chapter, you explored some of the most common numerical libraries.

First, I discussed algorithms based on matrix computation and how they can be represented using STL-based containers. The STL also provides a wealth of algorithms, which can be used in numerical applications as well as in other generic programming tasks. Next, you learned how to use the compile-time facilities provided by the C++ template mechanism. You have seen examples of how to employ such template-based facilities to calculate the factorial of a number. The same concepts can be extended for many other uses as well. You have also learned about the use of ratio templates and how they can represent financial concepts such as the Calmar ratio.

Probability distributions are another area of numerical algorithms that have a strong presence in financial applications. The testing of investment strategies usually involves the generation of stochastic data, as a way of simulating possible economic scenarios. You learned how to generate random values based on some of the most common probability distributions. Such distributions are provided by a few numeric libraries, and in this chapter, I have used boost::math and boost::random for this purpose. Together, these libraries provide a way to generate random data, as well as returning relevant information about specific distributions such as mean, standard deviation, quantiles, and other related attributes.

Data visualization is another area of programming that is very important in the development of effective financial algorithms. In the next chapter, you will explore a few programming techniques that exemplify some of the options available for data visualization. You will see that C++ has a lot of ways of outputting data to graphical displays, using both internal and external charting techniques. These libraries can be used to visualize every aspect of your work as you develop new investment strategies.

CHAPTER 6

Plotting Financial Data

A very common activity in financial programming is the generation of price-related data that needs to be visualized by traders or other business stakeholders. Most of the time, the data is expected to be plotted in the form of a chart for easy visualization. Visualization strategies for financial data range from simple line charts for daily prices to complex graphical output using candles, superposed studies, and other less conventional notation.

In this chapter, you will see a number of coding examples for creating and displaying charts based on prices and related quantitative data analysis. You will learn how to perform such tasks using a few different techniques, including external software such as Gnuplot as well as graphical C++ libraries such as Qt. Both techniques may be useful in different situations, as they have their own advantages and disadvantages.

The following are a few things you will learn in this chapter:

- How to create a class that provides a plotting interface

- How to use external plotting applications such as Gnuplot

- How to convert your data to a format that can be understood by external programs

- How to plot csv (comma-separated values) files on UNIX and Windows

- How to generate commands to control the open source Gnuplot application

- How to create a plot using an open source and multiplatform graphical user interface (GUI) library

- How to use Qt to generate a basic plotting window

© Carlos Oliveira 2021
C. Oliveira, *Practical C++20 Financial Programming*, https://doi.org/10.1007/978-1-4842-6834-6_6

Plotting with Gnuplot

Create a price chart using Gnuplot.

Solution

Gnuplot is a very popular software package used to create charts based on mathematical functions and data points. You can use Gnuplot in a stand-alone fashion or as an embedded viewer for graphs created by other applications. In this section, you will learn how to generate files that can be easily visualized using Gnuplot.

The first step in using Gnuplot is to make sure that it is properly installed in your system. You can easily install this package for data visualization by visiting its website (www.gnuplot.info) and downloading the required files. There are binary installation files available for most operating systems, including Windows, Mac OS X, and Linux. Run the installer and execute the main application. You should see something similar to the screen displayed in Figure 6-1.

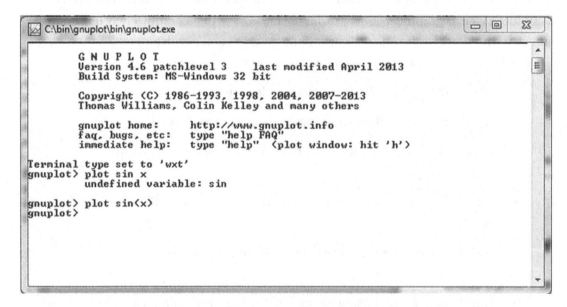

Figure 6-1. *Gnuplot main application running on Windows*

The basic application is composed of a simple shell where you can type some of the commands Gnuplot recognizes. The most basic of such commands is plot, which allows you to display plots on the screen. For example, you can easily create a plot for a mathematical function, such as sine or cosine. The command necessary for this can be typed at the main prompt of the application.

```
>  plot sin(x)
```

You can see the results for this simple function plot in Figure 6-2.

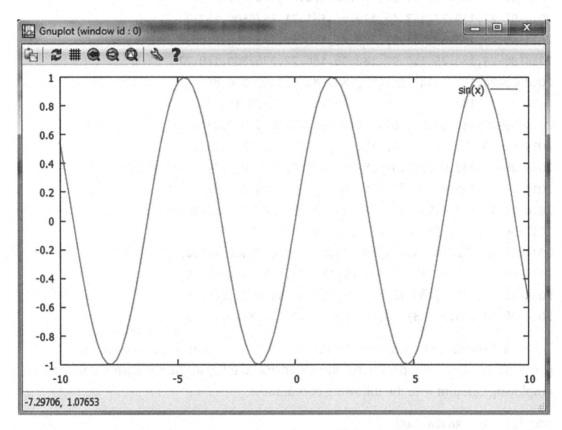

Figure 6-2. *Plot of the mathematical function sin(x) using Gnuplot*

In this plot, you supplied the mathematical function defined by `sin(x)`, and Gnuplot is responsible for creating a plot of the values, where the default range is from -10 to 10. You can easily tweak the parameters used to determine the range, as well as other attributes of the plot such as the title, the legend, and the units used in both axes.

Another way to use Gnuplot is to directly plot numeric data, instead of a mathematical function. This is possible by referencing the name of the files that should be imported by the plot command. Most data imported in this way is in the csv (comma-separated values) format, although Gnuplot doesn't mandate that the number be in csv—any file with numeric data organized as columns will do.

Consider the following data as an example. These are prices for IBM downloaded from Yahoo! Finance.

```
Date,Open,High,Low,Close,Volume,Adj Close
2014-07-01,181.7,187.27,181.7,186.35,6643100,186.35
2014-06-30,181.33,181.93,180.26,181.27,4223800,181.27
2014-06-27,179.77,182.46,179.66,181.71,4575500,181.71
2014-06-26,180.87,181.37,179.27,180.37,3258500,180.37
2014-06-25,180.25,180.97,180.06,180.72,2762800,180.72
2014-06-24,181.5,183,180.65,180.88,3875400,180.88
2014-06-23,181.92,182.25,181,182.14,3231700,182.14
2014-06-20,182.59,182.67,181.4,181.55,10686800,181.55
2014-06-19,184.12,184.47,182.36,182.82,3551100,182.82
2014-06-18,182.04,183.61,181.79,183.6,3931800,183.6
2014-06-17,181.9,182.81,181.56,182.26,2445400,182.26
2014-06-16,182.4,182.71,181.24,182.35,3538700,182.35
2014-06-13,182,183,181.52,182.56,2773600,182.56
2014-06-12,182.48,182.55,180.91,181.22,4425300,181.22
2014-06-11,183.61,184.2,182.01,182.25,4061700,182.25
2014-06-10,186.2,186.22,183.82,184.29,4154900,184.29
2014-06-09,186.22,187.64,185.96,186.22,2728400,186.22
```

I am displaying here only the few first lines of the file that contains daily stock prices. You can save this data in the file IBM.csv and use it as the source for a price plot employing Gnuplot with the following commands:

```
gnuplot> set xdata time
gnuplot> set datafile separator ","
gnuplot> set timefmt "%Y-%m-%d"
gnuplot> plot 'IBM.csv'  using 1:7  title columnhead with lines
```

Note   When running the previous commands, make sure that you're in the same directory in which you have saved the data file (IBM.csv). Another way to do this is to use the full path for the file, for example, "c:\\testdata\\IBM.csv" (escaped path separators are needed in the Windows platform).

The first command is used to tell Gnuplot that the data in the x axis is time-oriented. The second command defines the separator used in the file. The third command describes the date format stored in the csv file. Finally, the last line tells Gnuplot to plot the contents of file IBM.csv, using columns 1 and 7 (column 1 contains dates, while column 7 has adjusted closing prices), and with the title of the time series defined by the headers for each column of the csv file.

These commands will generate the plot displayed in Figure 6-3.

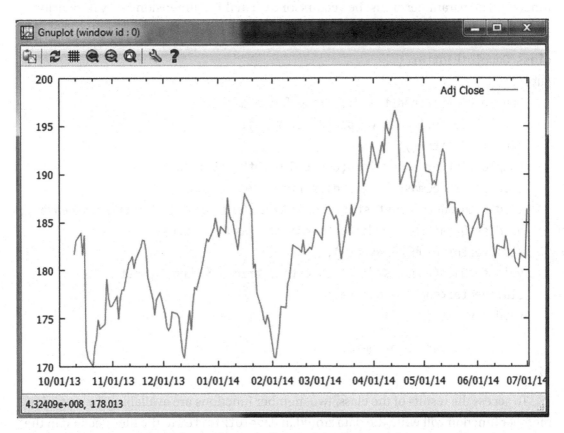

Figure 6-3. *Plot of the adjusted prices for IBM stock, stored in a csv file using Gnuplot*

The plot in Figure 6-3 is just a sample of what Gnuplot can do. There are literally hundreds of parameters that can be tweaked using the set command. Among these options, you can find three-dimensional plots, different colors, and line styles, among others.

To solve the problem presented in this section, you need to create a class that receives some data in the form of vectors of doubles or strings and produces output data suitable for consumption by Gnuplot. I have created such a class, which is called GnuplotPlotter and is responsible for the generation of the files needed by Gnuplot.

The operation of the class depends on the determination of data for the x axis as well as for the y axis. The class is created using a constructor that takes the output filename as a parameter. To define the data that will be used in the plot, use the setData member function. The parameters must be vectors for data in the x dimension and y dimension, respectively. The following is a summary of the class:

```
class GnuplotPlotter {
public:
    GnuplotPlotter(const std::string &fileName);
    GnuplotPlotter(const GnuplotPlotter &p);
    ~GnuplotPlotter();
    GnuplotPlotter &operator=(const GnuplotPlotter &p);
    void generateCmds(const std::string &cmdFileName);
    void setHeaders(const std::string &xheader, const std::string &yheader);
    void setData(const std::vector<double> &xdata, const
    std::vector<double> &ydata);
    void setData(const std::vector<std::string> &xdata, const
    std::vector<double> &ydata);
    void csvWrite();

    // private variables here.
};
```

To access the results of the class, two member functions are available. The csvWrite member function will write the data stored in GnuplotPlotter to the file specified in the constructor, using the csv format. The second member function is generateCmds, which allows one to create a command file with the necessary instructions to Gnuplot. This way, you don't need to worry about the exact syntax for plotting the file. The commands are stored in a filename specified by the parameter cmdFileName.

An example for the GnuplotPlotter class is given in the main function. First, you need to define two vectors with the desired data. In this case, you will use data generated by the function sin, which returns the trigonometric sine of a number. Notice that we do this only to simplify data testing. The data file, however, can have numbers generated from any source. After the content has been defined, you can call the member functions csvWrite and generateCmds to create the files needed by Gnuplot. You can see the result of this process in Figure 6-4.

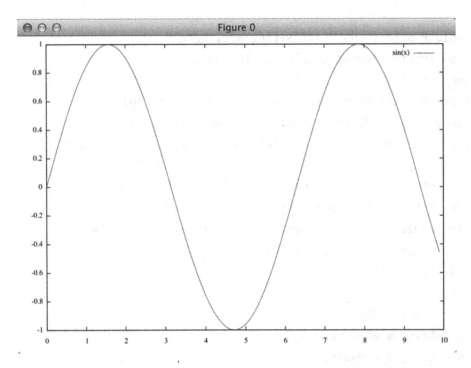

Figure 6-4. *Plot generated using the test data created in GnuplotPlotter.cpp*

Complete Code

The code to generate plots and Gnuplot commands has been implemented in the class GnuplotPlotter. You can add this class to your project and access the same member functions to generate data plots (see Listing 6-1).

Listing 6-1. GnuplotPlotter.h and GnuplotPlotter.cpp

```
//
// GnuplotPlotter.h

#ifndef __FinancialSamples__GnuplotPlotter__
#define __FinancialSamples__GnuplotPlotter__
```

```cpp
#include <vector>
#include <string>

class GnuplotPlotter {
public:
    GnuplotPlotter(const std::string &fileName);
    GnuplotPlotter(const GnuplotPlotter &p);
    ~GnuplotPlotter();
    GnuplotPlotter &operator=(const GnuplotPlotter &p);
    void generateCmds(const std::string &cmdFileName);
    void setHeaders(const std::string &xheader, const std::string
    &yheader);
    void setData(const std::vector<double> &xdata, const
    std::vector<double> &ydata);
    void setData(const std::vector<std::string> &xdata, const
    std::vector<double> &ydata);
    void csvWrite();
private:
    std::string m_fileName;
    std::string m_xheader;
    std::string m_yheader;
    std::vector<std::string> m_xdata;
    std::vector<double> m_ydata;
    bool m_isDate;
};

#endif /* defined(__FinancialSamples__GnuplotPlotter__) */

//
//  GnuplotPlotter.cpp

#include "GnuplotPlotter.h"

#include <fstream>
#include <iostream>
#include <sstream>
#include <cmath>
```

```
using std::ofstream;
using std::vector;
using std::cout;

GnuplotPlotter::GnuplotPlotter(const std::string &fileName)
: m_fileName(fileName),
  m_isDate(false)
{
}

GnuplotPlotter::GnuplotPlotter(const GnuplotPlotter &p)
: m_fileName(p.m_fileName),
  m_xheader(p.m_xheader),
  m_yheader(p.m_yheader),
  m_xdata(p.m_xdata),
  m_ydata(p.m_ydata),
  m_isDate(p.m_isDate)
{
}

GnuplotPlotter::~GnuplotPlotter()
{
}

GnuplotPlotter &GnuplotPlotter::operator=(const GnuplotPlotter &p)
{
        if (&p != this)
        {
                m_fileName = p.m_fileName;
                m_xheader = p.m_xheader;
                m_yheader = p.m_yheader;
                m_xdata = p.m_xdata;
                m_ydata = p.m_ydata;
                m_isDate = p.m_isDate;
        }
    return *this;
}
```

```cpp
void GnuplotPlotter::setData(const std::vector<std::string> &xdata,
                             const std::vector<double> &ydata)
{
        m_xdata = xdata;
        m_ydata = ydata;
        m_isDate = true; // assume that x-axis is a date
}

void GnuplotPlotter::setData(const std::vector<double> &xdata, const
std::vector<double> &ydata)
{
        for (unsigned i = 0; i < xdata.size(); ++i)
        {
                std::stringstream ss;
                ss << xdata[i];
                m_xdata.push_back(ss.str());
        }
        m_ydata = ydata;
        m_isDate = false; // x-axis cannot be a date.
}

void GnuplotPlotter::setHeaders(const std::string &xheader, const
std::string &yheader)
{
        m_xheader = xheader;
        m_yheader = yheader;
}

void GnuplotPlotter::generateCmds(const std::string &cmdFileName)
{
        ofstream file;

        file.open(cmdFileName.c_str());
        if (file.fail())
        {
                cout << "failed to open file " << m_fileName << endl;
                return;
        }
```

```
        if (m_isDate)
        {
                file << "set xdata time"  << endl;
                file << "set timefmt \"%Y-%m-%d\" "  << endl;
        }
        file << "set datafile separator \",\" "  << endl;
        file << "plot '" << m_fileName <<   "'  u 1:7  title columnhead
        w lines " << endl;
        file << "pause -1" << endl;
}

void GnuplotPlotter::csvWrite()
{
        ofstream file;

        file.open(m_fileName.c_str());
        if (file.fail())
        {
                cout << "failed to open file " << m_fileName << endl;
                return;
        }

        if (m_xdata.size() != m_ydata.size())
        {
                cout << "data has incorrect size " << endl;
                return;
        }

        file << m_xheader << "," << m_yheader << endl;

        for (unsigned i = 0; i < m_xdata.size(); ++i)
        {
                file << m_xdata[i] << "," << m_ydata[i] << endl;
        }
}
```

```cpp
int main()
{
    GnuplotPlotter plotter("test.csv");
    plotter.setHeaders("x", "sin(x)");

    vector<double> xdata;
    vector<double> ydata;
    for (int i=0; i<100; ++i)
    {
        double x = i*10/100.0;
        xdata.push_back(x);
        ydata.push_back(sin(x));
    }

    plotter.setData(xdata, ydata);
    plotter.csvWrite();
    plotter.generateCmds("testcmds.gp");
    return 0;
}
```

Running the Code

The code in Listing 6-1 can be compiled using the free gcc compiler. The solution was tested on the Mac OS X and Windows platforms. You can, for example, create an application using the following command:

```
gcc -o gnuplotter gnuplotplotter.cpp
```

Then, you can run the program using the command line

```
./gnuplotter
```

This will generate two files, test.csv and testcmds.gp, which Gnuplot will use to generate the desired plot. You can run Gnuplot on UNIX as follows:

```
cat testcmds.gp | gnuplot
```

In the Windows platform, you can load the commands file into the Gnuplot application in the following way:

```
c:> gnuplot
> load "testcmds.gp"
```

The plot will be displayed in a separate window, as shown in Figure 6-4.

Plotting Data from a GUI

Create an application that can plot data using the GUI.

Solution

Although it is great to have the ability to create charts with external packages such as Gnuplot, sometimes it is necessary to have a larger degree of control over the output generated by plots. If you cannot find a way to use one of the parameters in Gnuplot to get the desired results, it becomes necessary to implement a plotting solution that runs in C++. This section shows how to achieve this.

There are many graphical libraries available for C++ developers, and the final decision depends mostly on your target environment. However, in this section, I use the Qt library to implement the desired solution.

Qt is probably the easiest to use graphical programming package around. You will see that with just a dozen lines, we are able to create a complete application. Moreover, Qt is available for all major operating systems, so that your application can be easily ported to other targets as necessary.

The class used is called QtPlotter, and it receives data using the setData member function, just as we did with the GnuplotPlotter. The main part of the implementation, however, is performed in the PlotWindow class, which is derived from QMainWindow, one of the key classes in the Qt framework. The PlotWindow class is responsible for managing the window and, most important, painting the plot when necessary.

The plotting functionality is implemented in the paintEvent member function. This member function is invoked whenever the window needs to paint itself. First, it paints the x and y axis and calculates the size of a unit on each axis, storing that information in variables called unitX and unitY. To draw the axis, the paintEvent member function uses the painter object, which is provided by Qt. The drawLine member function is the simplest way to draw line between the given coordinates, as shown in the following code:

```
// define margins
double marginX = 10;
double marginY = 10;
double lengthX = 500;
double lengthY = 400;

// define axis
int maxX = lengthX, maxY = lengthY;
painter.drawLine(marginX,marginY, marginX, lengthY+marginY);
painter.drawLine(marginX,lengthY + marginY, lengthX, lengthY + marginY);
```

In the next step, the function paintEvent draws the tick markers along the axis. Finally, the code paints lines between the points given as input to the plot.

The last part of the implementation is encapsulated in the plotWindowRun member function, which is part of the QtPlotter class, as follows:

```
int QtPlotter::plotWindowRun()
{
    char *arg = (char *)"plotter";
    int argc = 1;
    QApplication app(argc, &arg);

    app.setApplicationName("Qt Plotter");
    PlotWindow window;

    window.resize(600, 600);
    window.show();
    return app.exec();
}
```

This code does most of what is necessary to create a Qt application and display a window on the screen. The window created is the PlotWindow class that we discussed previously, so that the plot is displayed as desired. The QApplication object is part of the Qt framework. It manages the workflow of a graphical application, including menus and windows. When creating a `QApplication`, we are able to determine the application name with the `setApplicationName` member function. Finally, we resize and show the plot window and call the `exec` member function to start the window display loop.

Figure 6-5 shows the results of this code.

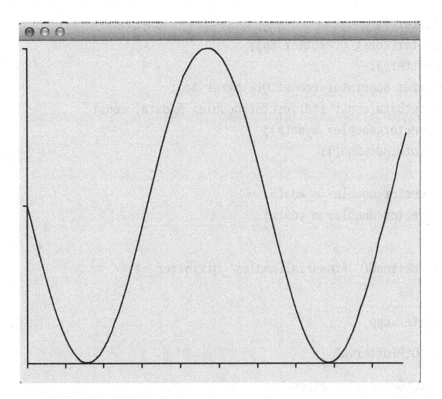

Figure 6-5. *Plot produced by class QtPlotter*

Complete Code

The class `QtPlotter`, displayed in Listing 6-2, implements the necessary functionality to show a plot in a Qt window, as explained in the previous section. To compile this code, you need to install the Qt libraries in your system.

Listing 6-2. QtPlotter.h and QtPlotter.cpp

```
//
// QtPlotter.h

#ifndef __FinancialSamples__QtPlotter__
#define __FinancialSamples__QtPlotter__
```

```cpp
#include <string>
#include <vector>

class QtPlotter {
public:
    QtPlotter();
    QtPlotter(const QtPlotter &p);
    ~QtPlotter();
    QtPlotter &operator=(const QtPlotter &p);
    void setData(const std::vector<double> &xdata, const
    std::vector<double> &ydata);
    int plotWindowRun();
private:
    std::vector<double> m_xdata;
    std::vector<double> m_ydata;
};

#endif /* defined(__FinancialSamples__QtPlotter__) */

//
//   QtPlotter.cpp

#include "QtPlotter.h"

#include <QtGui/qapplication.h>
#include <QtGui/qmainwindow.h>
#include <QtGui/qpainter.h>

#include <algorithm>
#include <cmath>

#include <iostream>

using std::vector;

class PlotWindow : public QMainWindow {
public:
    PlotWindow();
    ~PlotWindow();
    void paintEvent(QPaintEvent *event);
```

```cpp
    void setData(const vector<double> &xdata, const vector<double> &ydata);
private:
    vector<double> m_xdata;
    vector<double> m_ydata;
};

PlotWindow::PlotWindow()
{
}

PlotWindow::~PlotWindow()
{
}

void PlotWindow::setData(const vector<double> &xdata, const vector<double>
&ydata)
{
    m_xdata = xdata;
    m_ydata = ydata;
}

void PlotWindow::paintEvent(QPaintEvent *event)
{
    QMainWindow::paintEvent(event);
    QPainter painter(this);

    // define margins
    double marginX = 10;
    double marginY = 10;
    double lengthX = 500;
    double lengthY = 400;

    // define axis
    int maxX = lengthX, maxY = lengthY;
    painter.drawLine(marginX,marginY, marginX, lengthY+marginY);
    painter.drawLine(marginX,lengthY + marginY, lengthX, lengthY + marginY);
```

```cpp
    // find units
    int largeX = 0, largeY = 0;
    double largeXd = 0, largeYd = 0;
    for (unsigned i=1; i<m_xdata.size(); ++i)
    {
        if (largeXd < m_xdata[i]) largeXd = m_xdata[i];
        if (largeYd < m_ydata[i]) largeYd = m_ydata[i];
    }
    largeX = (int)largeXd + 1;
    largeY = (int)largeYd + 1;

    int unitX = maxX / largeX;
    int unitY = maxY / largeY;

    // paint ticks
    for (int i=0; i<largeY; ++i)
    {
        painter.drawLine(marginX-5, i*unitY+marginY, marginX,
        i*unitY+marginY);
    }
    for (int i=0; i<largeX; ++i)
    {
        painter.drawLine(marginX+i*unitX, lengthY+marginY, marginX+i*unitX,
        lengthY+5+marginY);
    }

    // draw plot
    for (unsigned i=1; i<m_xdata.size(); ++i)
    {
        painter.drawLine(marginX+unitX*m_xdata[i-1], unitY*m_ydata[i-1]+
        marginY,
                        marginX+unitX*m_xdata[i], unitY*m_ydata[i]+marginY);
    }
}

QtPlotter::QtPlotter()
{
}
```

```
QtPlotter::~QtPlotter()
{
}

QtPlotter::QtPlotter(const QtPlotter&p)
: m_xdata(p.m_xdata),
  m_ydata(p.m_ydata)
{

}

QtPlotter &QtPlotter::operator=(const QtPlotter &p)
{
    if (&p != this)
    {
        m_xdata = p.m_xdata;
        m_ydata = p.m_ydata;
    }
    return *this;
}

void QtPlotter::setData(const std::vector<double> &xdata, const
std::vector<double> &ydata)
{
    m_xdata = xdata;
    m_ydata = ydata;
}

int QtPlotter::plotWindowRun()
{
    char *arg = (char *)"plotter";
    int argc = 1;
    QApplication app(argc, &arg);

    app.setApplicationName("Qt Plotter");
    PlotWindow window;
```

```
    window.resize(600, 600);
    window.show();
    return app.exec();
}

int main()
{
    QtPlotter plotter;

    vector<double> xdata;
    vector<double> ydata;
    for (int i=0; i<100; ++i)
    {
        double x = i*10/100.0;
        xdata.push_back(x);
        ydata.push_back(sin(x)+1);
    }

    plotter.setData(xdata, ydata);

    return plotter.plotWindowRun();
}
```

Running the Code

To use the QtPlotter class, you need to have the Qt4 library installed in your system. The installation process requires you to visit the developer website (`www.qt.io/developers/`), download, and run the installer application. After the installation is complete, the libraries will be copied to a user-defined folder.

The next step is to tell your compiler or IDE (Integrated Development Environment) where the libraries can be found. The two main parameters are the include path (used by the compiler) and the link path (used by the linker). For example, if Qt4 was installed in the directory `/usr/local/qt4`, the include path should be `/usr/local/qt4/include`, and the link path should be `/usr/local/qt4/lib`. From the library directory, at least two libraries are needed: `libQtCore` and `libQtGui`. You can refer to the Qt documentation for details on how to link to Qt libraries for specific systems, such as Windows. To compile and link your application using gcc, for example, the following command line would provide the necessary information:

```
$ gcc -o qtExample QtPlotter.cpp -I/usr/local/qt4/include -L/usr/local/qt4/
lib -lQtGui -lQtCore
```

Conclusion

In this chapter, you learned a few techniques to plot financial data using C++. Visualization is one of the factors that shouldn't be overlooked in the creation of efficient investment strategies. The better your visualization facilities, the easier it is to spot trends and opportunities in the markets. While there are many free and commercial alternatives to display stock charts, we frequently need to present data in a more flexible way.

I started the chapter with a recipe for creating numerical plots using Gnuplot. Gnuplot is a free, widely available package for data visualization, which runs in most operating systems, including Windows, Mac OS X, and UNIX. You have seen how to create a class that encapsulates the information necessary to create graphs in Gnuplot.

The next section gave you another approach to create your own financial plots, using a C++ graphical library called Qt. You can employ this type of code in multiple platforms, taking advantage of the high portability of the underlying framework. The QtPlotter class presented here exposes an interface that your program can use to display a single plot based on values for the x and y axis.

Many of the algorithms in finance depend on the solution of systems of equations, which are based on linear algebra concepts. For the developer on the financial industry, it is very useful to have a basic knowledge of linear algebra and its software implementations. These concepts can be viewed as the building blocks used by financial engineers and can be easily accessed in C++. In the next chapter, you will see a few programming examples that make use of linear algebra concepts as part of financial applications.

CHAPTER 7

Linear Algebra

Linear algebra is a fundamental set of mathematical tools that has applications in many areas of science and engineering. Consequently, linear algebra (LA) techniques also play an important role in the practice of financial programming, and they are frequently used throughout the area of financial engineering. LA-based techniques are frequently used in the development of trading strategies.

As C++ programmers, it is important to understand how the traditional methods of linear algebra can be integrated in financial applications. With this goal in mind, I present a few examples that show how to use some of the most common LA algorithms along with other C++ libraries. In this chapter, you will also learn how to integrate existing LA libraries into your code, with special attention to the uBLAS library included with boost.

The following are some topics that we cover in this chapter:

- Basic operations of linear algebra: You will learn how to use the fundamental functions of linear algebra in your code.

- BLAS (basic linear algebra subprograms) library overview: BLAS is a well-known set of functions that are nowadays the standard for LA implementations. You will learn about the three levels of BLAS support along with their functionality.

- uBLAS: BLAS is function-based library, which has been used in languages such as Fortran and C. To use the higher-level concepts of modern C++, the boost project has created a new implementation. You will learn how the uBLAS library implements the same concepts present on BLAS.

- Computing determinants: Calculating the determinant of a matrix is one of the most common tasks when analyzing a set of linear equations. You will learn how to use uBLAS to perform this type of LA computation.

C. Oliveira, *Practical C++20 Financial Programming*, https://doi.org/10.1007/978-1-4842-6834-6_7

- Converting between standard types and uBLAS types: You will see how to convert standard types such as std::vector into types that are more appropriate for LA computations.

Using Basic Linear Algebra Operations

Create a class that performs basic LA operations such as vector and scalar products.

Solution

Linear algebra has been used to solve a large number of engineering and scientific problems. As such, these concepts are frequently employed as part of financial applications. The basic level of computational linear algebra deals with scalars and vector and with the operations allowed on these mathematical entities.

A scalar is a quantity that is composed of a single measurement. Normally, it doesn't require the creation of a separate class, since it can be easily represented as an integer, a floating point, or a double number. Such quantities are also usually stored as a single element. Scalar numbers enjoy the associated traditional properties such as addition, subtraction, multiplication, and division.

The use of scalars needs no special treatment in C++ implementations, although certain classes may treat scalar parameters as a template argument, so that you can later work with different types. For example, the following is common on numeric libraries:

```
template <class Scalar>
class MyNumericClass {
    void aFunction(Scalar parameter);

public:
    Scalar m_internalVar;
};
```

In this case, the Scalar type acts as a placeholder for one of the types supported by C++, such as int, float, or long double. In this way, you can easily parameterize the numeric class according to the actual type needed for the computation and avoid the unnecessary cast between numeric types, which can introduce unexpected errors to the computation.

The next level of LA operations includes the combination of vectors, using vector addition, vector product, and scalar multiplication. Initially, you may think about employing `std::vector` to perform such operations; however, `std::vector` is a general-purpose container that is not tuned for mathematical processing.

The traditional solution for LA implementation is using the BLAS (basic linear algebra subprograms) library. BLAS is a popular package that was originally implemented in Fortran but has since then become the standard for LA computation even for other languages. The C-language version was created with the use of the f2c converter from Fortran. Since many LA packages came to rely on the functionality provided by BLAS, other libraries have been created to emulate it whenever necessary.

In this section, I introduce a C++ library that implements much of the functionality of BLAS. The uBLAS library is part of boost and can be accessed by including one of the header files such as `<boost/numeric/ublas/vector.hpp>`.

BLAS and similar libraries are organized according to support levels, ranging from 1 to 3. The BLAS support levels include the following:

- Level 1: Support for operations using scalar numbers and vectors. At this level, the library offers support for numeric vectors in one dimension, with common operations such as scalar multiplication and vector product.

- Level 2: At the second level, BLAS-compatible libraries provide functions to perform computations involving vectors and matrices— for example, the common multiplication of a vector v by a matrix A, which can be performed (with different results) as with vA or Av.

- Level 3: The third level of BLAS is defined for matrix-matrix operations. It allows, for example, the multiplication of matrices.

These three levels of BLAS support have been implemented in several libraries inspired in the original BLAS. Such implementations are mostly created to effectively support new programming languages, architectures, and processors while still maintaining compatibility with the many numeric algorithms that depend on BLAS. The purpose of a boost uBLAS library is to provide the same support levels of BLAS while taking advantage of the expressive power provided by C++ classes and templates.

In this example, you will explore a class called `VectorOperations`, which is responsible for implementing level 1 BLAS operations. This means that it has to deal with vectors and scalar numbers, as well as the possible transformations allowed between them. From the documentation of BLAS, we have the following categories of operations:

- Swap: Switches element from the first vector to the second vector.

- Scale: Multiplies all elements of a vector by a single scalar number.

- Copy: Performs a copy of elements of a first vector into a second, destination vector.

- Vector addition: Returns a vector whose components are the element-wise additions of two input vectors.

- Dot product: Performs the mathematical operation of inner vector product, which is defined for two vectors v and w using the following formula:

$$p(v,w) = \sum_i (v_i + w_i)$$

- Norm: The norm of a vector is a way to quantify the length of a vector in a particular direction. A common norm is the two-dimensional distance between two points.

In the implementation of VectorOperations, you will see how some of these operations can be accessed using uBLAS. The first such operation is vector multiplication by a scalar. The method signature is as follows:

```
std::vector<double> scalarMult(double scalar);
```

The goal of this method is to return a std::vector object where each member is a scaled version of the elements in the original vector. The implementation shows how to convert between these different vector types.

```
std::vector<double> VectorOperations::scalarMult(double scalar)
{
    using namespace boost::numeric::ublas;
    vector<double> vx;

    std::copy(m_data.begin(), m_data.end(), vx.end());
```

```
    vector<double> res = vx * scalar;

    std::vector<double> v;
    std::copy(res.begin(), res.end(), v.end());

    return v;
}
```

The first step is to create a vector from the `boost::numeric::ublas` namespace. Notice that this function employs the `using` declaration to avoid the boring sequence of namespaces. The next step is to make a copy from the original `std::vector` into the ublas vector. Finally, the scalar operation is performed using the multiplication operator. To store the result, a new `ublas` vector, called `res`, is constructed. The last step is to copy the result into a new vector and return the result.

The previous algorithm creates a lot of temporaries, and therefore it is not efficient for real implementations. However, the fact that we convert from standard vectors to ublas vectors has the advantage of bringing attention to what each vector type is capable of doing. When implementing more complicated LA algorithms, however, we should avoid the creation of any unnecessary temporary variables, since they can occupy a lot of space for large vectors and matrices.

The `VectorOperations` class presents similar examples for other common operations you will find on BLAS level 1, such as `addVector` and `subtractVector`, which use the `ublas` operators to quickly perform these computations. The `dotProduct` member function uses the `inner_prod` function from `ublas` to implement the dot product, also known as inner product operation between vectors. Finally, we have the example of the `norm` member function, which returns the length of the vector as defined with the `norm_2` function.

Complete Code

The vector operations described in the previous section have been implemented in the `VectorOperations` class, displayed in Listing 7-1. You should be able to compile the class using any standards-compliant C++ compiler, after you install the boost libraries in your system.

Listing 7-1. VectorOperations.h and VectorOperations.cpp

```
//
//  VectorOperations.h

#ifndef __FinancialSamples__VectorOperations__
#define __FinancialSamples__VectorOperations__

#include <vector>

// performs operations on std::vector using boost ublas
class VectorOperations {
public:
    VectorOperations(const std::vector<double> &v);
    VectorOperations(const VectorOperations &p);
    ~VectorOperations();
    VectorOperations &operator=(const VectorOperations &p);
    std::vector<double> scalarMult(double scalar);
    std::vector<double> addVector(const std::vector<double> &v);
    std::vector<double> subtractVector(const std::vector<double> &v);
    double dotProd(const std::vector<double> &v);
    double norm();
private:
    std::vector<double> m_data;
};
#endif /* defined(__FinancialSamples__VectorOperations__) */

//
//  VectorOperations.cpp

#include "VectorOperations.h"

#include <boost/numeric/ublas/vector.hpp>

VectorOperations::VectorOperations(const std::vector<double> &p)
: m_data(p)
{
}
```

```cpp
VectorOperations::VectorOperations(const VectorOperations &p)
: m_data(p.m_data)
{

}

VectorOperations::~VectorOperations()
{
}

VectorOperations &VectorOperations::operator=(const VectorOperations &p)
{
    if (this != &p)
    {
        m_data = p.m_data;
    }
    return *this;
}

std::vector<double> VectorOperations::scalarMult(double scalar)
{
    using namespace boost::numeric::ublas;
    vector<double> vx;

    std::copy(m_data.begin(), m_data.end(), vx.end());

    vector<double> res = vx * scalar;

    std::vector<double> v;
    std::copy(res.begin(), res.end(), v.end());

    return v;
}

std::vector<double> VectorOperations::addVector(const std::vector<double>
&vec)
{
    using namespace boost::numeric::ublas;
```

```cpp
    vector<double> v1;
    std::copy(m_data.begin(), m_data.end(), v1.end());

    vector<double> v2;
    std::copy(vec.begin(), vec.end(), v2.end());

    vector<double> v3 = v1 + v2;

    std::vector<double> v;
    std::copy(v3.begin(), v3.end(), v.end());
    return v;
}

double VectorOperations::norm()
{
    using namespace boost::numeric::ublas;

    vector<double> v1;
    std::copy(m_data.begin(), m_data.end(), v1.end());

    double res = norm_2(v1);
    return res;
}

std::vector<double> VectorOperations::subtractVector(const
std::vector<double> &vec)
{
    using namespace boost::numeric::ublas;

    vector<double> v1;
    std::copy(m_data.begin(), m_data.end(), v1.end());

    vector<double> v2;
    std::copy(vec.begin(), vec.end(), v2.end());

    vector<double> v3 = v1 - v2;

    std::vector<double> v;
    std::copy(v3.begin(), v3.end(), v.end());
    return v;
}
```

```
double VectorOperations::dotProd(const std::vector<double> &v)
{
    using namespace boost::numeric::ublas;

    vector<double> v1;
    std::copy(m_data.begin(), m_data.end(), v1.end());

    vector<double> v2;
    std::copy(v.begin(), v.end(), v2.end());

    double res = inner_prod(v1, v2);
    return res;
}
```

Using Matrix-Oriented Operations

In this section, we create a class to perform matrix operations compatible with BLAS.

Solution

As you learned from Listing 7-1, LA functions are designed to work with linear operators that use scalar numbers, vectors, and matrices. To support these operations, programmers use a set of functions that are compatible with the original BLAS library. In C++, we can have access to a few libraries that implement BLAS, including the uBLAS library from boost, which you have been using so far.

The second level of BLAS is responsible for providing support for matrix-vector operations. In this example, you will see how to implement functions that use this level of BLAS. You will use boost uBLAS to access this functionality.

At level 2 of BLAS, the goal is to allow for the combination of vectors and matrices. To make this possible, uBLAS implements a few higher-level classes that incorporate concepts defined in the BLAS framework. The following are some of the most important classes used:

- Vector: This class was already discussed in the last section, and it acts as a general container for vector data. Some other classes implement restrictions on the generic functionality of vector.

- Sparse vector: A specialized version of a vector that allows for data represented in a sparse way. It can be used whenever the number of nonzero elements in a vector is small compared to the size of the array.

- Matrix: This is the main class that represents a two-dimensional arrangement of values, which is the traditional representation of a matrix.

- Triangular matrix: This class is used to represent matrices in which data is stored only on or above the main diagonal (for upper triangular matrices). You can also create lower triangular matrices with uBLAS.

- Symmetric matrix: This type of matrix has elements that are symmetric with respect to the diagonal. This class is used to represent this type of matrix on algorithms that take advantage of this property.

- Hermitian matrix: A Hermitian matrix has the property that its elements are complex numbers, and there is a symmetry based on the notion of complex conjugate. That is, for each entry at position [i,j], the corresponding elements [j,i] are its complex conjugate.

- Banded matrix: This class represents sparse matrices where the nonzero elements are stored in a narrow band of elements around the main diagonal. The size of the band can be specified when creating the matrix.

- Sparse matrix: A class that represents a generic sparse matrix—that is, a matrix where most elements are zero. By using a sparse matrix, you can avoid the need to store in memory a large number of zero values.

Using these classes, you can easily store the data using the best representation available for the required task. Using the right representation can also give you a great advantage in finding the right algorithm, since uBLAS automatically provides special versions of its operators based on the data types used. For example, if a matrix is known to be triangular, it is possible to speed up some computations when solving a system of equations. This means that using a `TriangularMatrix` instead of a generic `Matrix` can result in a substantial speedup for your code.

To explore the operations available for matrices, I introduce a class called
`MatrixOperations`. This class is able to convert parameters into the classes required by
uBLAS. It is also responsible for calling uBLAS operators on these converted parameters.
Using this class, you have an easy way to test several of the functions that operate on
matrix parameters.

Among the main matrix-related functions and operators in uBLAS, you will find the
following:

- Scalar multiplication: Multiplying a matrix by a scalar is a simple
  process, since it uses the standard multiplication operator in C++.
  You just need to save the result of the multiplication in a new matrix
  variable.

- Vector multiplication: Multiplying a matrix by a vector is a common
  operation. You can do this using the `prod` function. The product can
  be performed in two ways: a pre-multiplication requires that the
  vector be the first argument for the `prod` function. You can also post-
  multiply by a vector, in which case the vector enters as the second
  parameter to the `prod` function.

- Matrix multiplication: You can also multiply two matrices. This
  results in a third matrix, which has a size defined by the sizes of the
  two original matrices. You can perform the multiplication operation
  using an overridden version of the `prod` function.

- Element-wise multiplication: This operation performs the
  multiplication of each corresponding element in the matrix. That
  is, given matrices A and B, the resulting matrix C is composed of
  elements `C[i,j] = A[i,j] + B[i,j]`.

- `Transposition`: The transpose of a matrix is a simple operation
  where elements `A[i,j]` and `A[j,i]` are exchanged. This results in
  a matrix that is the transposition of the original around the main
  diagonal.

In the `MatrixOperations` class, you will find examples of each of these operations.
The arguments and return values for the member functions of `MatrixOperations`
are given in terms of standard vectors (`std::vector`) or `Matrix` objects (which you

learned about in Chapter 5). While this kind of conversion should be avoided in high-performance code, you can take it as an example of what is necessary to create objects of the types declared in uBLAS. Consider, for instance, the method transpose.

```
Matrix MatrixOperations::transpose()
{
        using namespace ublas;
        int d1 = m_rows.size();
        int d2 = m_rows[0].size();
        matrix<double> M(d1, d2);

        for (int i = 0; i < d1; ++i)
        {
                for (int j = 0; j < d2; ++j)
                {
                        M(i,j) = m_rows[i][j];
                }
        }

        matrix<double> mp = trans(M);
        return fromMatrix(mp);
}
```

The first step is to determine the size of the matrix you need to build, which is given by the dimensions d1 and d2. Using this information, you can create a new ublas::matrix object. You will then initialize the matrix using the data stored in the m_rows member variable. Finally, you can call the trans function from uBLAS, which is responsible for doing the transpose of its argument. The last step is to convert from the uBLAS representation to a Matrix object, which is performed by the fromMatrix function.

Complete Code

Listing 7-2 shows the implementation of the class MatrixOperations. The code makes use of the Matrix class implemented in Chapter 5, so you need to add it to your compilation line.

Listing 7-2. MatrixOperations.h and MatrixOperations.cpp

```
//
// MatrixOperations.h

#ifndef __FinancialSamples__MatrixOperations__
#define __FinancialSamples__MatrixOperations__

#include <vector>

#include "Matrix.h"

class MatrixOperations {
public:
    MatrixOperations();
    ~MatrixOperations();
    MatrixOperations(const MatrixOperations &p);
    MatrixOperations &operator=(const MatrixOperations &p);

    void addRow(const std::vector<double> &row);
    Matrix multiply(Matrix &m);
    Matrix transpose();
    Matrix elementwiseMultiply(Matrix &m);
    Matrix scalarMultiply(double scalar);
    std::vector<double> preMultiply(const std::vector<double> &v);
    std::vector<double> postMultiply(const std::vector<double> &v);

private:
    std::vector<std::vector<double> > m_rows;
};

#endif /* defined(__FinancialSamples__MatrixOperations__) */

//
// MatrixOperations.cpp

#include "MatrixOperations.h"

#include <boost/numeric/ublas/matrix.hpp>
#include <boost/numeric/ublas/io.hpp>
#include <boost/numeric/ublas/lu.hpp>
```

```cpp
namespace ublas = boost::numeric::ublas;
using std::cout;
using std::endl;

MatrixOperations::MatrixOperations()
{
}

MatrixOperations::~MatrixOperations()
{
}

void MatrixOperations::addRow(const std::vector<double> &row)
{
        m_rows.push_back(row);
}

static Matrix fromMatrix(const ublas::matrix<double> &mp)
{
        using namespace ublas;

        int d1 = mp.size1();
        int d2 = mp.size2();
        Matrix res(d1, d2);
        for (int i = 0; i < d1; ++i)
        {
                for (int j = 0; j < d2; ++j)
                {
                        res[i][j] = mp(i,j);
                }
        }
        return res;
}

Matrix MatrixOperations::elementwiseMultiply(Matrix &m)
{
        using namespace ublas;
        int d1 = m_rows.size();
```

```
        int d2 = m_rows[0].size();
        matrix<double> M(d1, d2);

        for (int i = 0; i < d1; ++i)
        {
                for (int j = 0; j < d2; ++j)
                {
                        M(i,j) = m_rows[i][j];
                }
        }

        matrix<double> M2(d1, d2);
        for (int i = 0; i < d1; ++i)
        {
                for (int j = 0; j < d2; ++j)
                {
                        M2(i,j) = m[i][j];
                }
        }

        matrix<double> mp = element_prod(M, M2);
        return fromMatrix(mp);
}

Matrix MatrixOperations::transpose()
{
        using namespace ublas;
        int d1 = m_rows.size();
        int d2 = m_rows[0].size();
        matrix<double> M(d1, d2);

        for (int i = 0; i < d1; ++i)
        {
                for (int j = 0; j < d2; ++j)
                {
                        M(i,j) = m_rows[i][j];
                }
        }
```

```
        matrix<double> mp = trans(M);
        return fromMatrix(mp);
}

Matrix MatrixOperations::multiply(Matrix &m)
{
        using namespace ublas;
        int d1 = m_rows.size();
        int d2 = m_rows[0].size();
        matrix<double> M(d1, d2);

        for (int i = 0; i < d1; ++i)
        {
                for (int j = 0; j < d2; ++j)
                {
                        M(i,j) = m_rows[i][j];
                }
        }

        matrix<double> M2(d1, d2);
        for (int i = 0; i < d1; ++i)
        {
                for (int j = 0; j < d2; ++j)
                {
                        M2(i,j) = m[i][j];
                }
        }

    matrix<double> mp = prod(M, M2);
    return fromMatrix(mp);
}

Matrix MatrixOperations::scalarMultiply(double scalar)
{
        using namespace ublas;
        int d1 = m_rows.size();
        int d2 = m_rows[0].size();
        matrix<double> M(d1, d2);
```

```
        for (int i = 0; i < d1; ++i)
        {
                for (int j = 0; j < d2; ++j)
                {
                        M(i,j) = m_rows[i][j];
                }
        }

        matrix<double> mp = scalar * M;
        return fromMatrix(mp);
}

std::vector<double> MatrixOperations::preMultiply(const std::vector<double> &v)
{
        using namespace ublas;
        ublas::vector<double> vec;
        std::copy(v.begin(), v.end(), vec.end());

        int d1 = m_rows.size();
        int d2 = m_rows[0].size();
        ublas::matrix<double> M(d1, d2);

        for (int i = 0; i < d1; ++i)
        {
                for (int j = 0; j < d2; ++j)
                {
                        M(i,j) = m_rows[i][j];
                }
        }

        vector<double> pv = prod(vec, M);

        std::vector<double> res;
        std::copy(pv.begin(), pv.end(), res.end());
        return res;
}
```

```cpp
std::vector<double> MatrixOperations::postMultiply(const
std::vector<double> &v)
{
        using namespace ublas;
        ublas::vector<double> vec;
        std::copy(v.begin(), v.end(), vec.end());

        int d1 = m_rows.size();
        int d2 = m_rows[0].size();
        ublas::matrix<double> M(d1, d2);

        for (int i = 0; i < d1; ++i)
        {
                for (int j = 0; j < d2; ++j)
                {
                        M(i,j) = m_rows[i][j];
                }
        }

        vector<double> pv = prod(M, vec);

        std::vector<double> res;
        std::copy(pv.begin(), pv.end(), res.end());
        return res;
}

int main()
{
    MatrixOperations op;
    for (int i=0; i<5; ++i)
    {
        std::vector<double> row;
        for (int j=0; j<5; ++j)
        {
            row.push_back(sin((double)j+i));
        }
        op.addRow(row);
    }
```

```
    op.transpose();
    Matrix res = op.scalarMultiply(12);
    return 0;
}
```

Running the Application

The code shown in Listing 7-2 can be compiled using any standards-conforming C++ compiler. You need to have boost installed in your system to access uBLAS (I used version 1.55, tested on Windows MingW and Mac OS X). For example, using the gcc compiler on a UNIX system can be done with the following command:

```
gcc -o matrixOp matrixOperations.cpp
```

This will result in an application called matrixOp. You can run the resulting application as

```
./matrixOp
```

This will run the test main function, which should print out the result of the requested operations. In my system, I got the following results:

```
0 10.0977 10.9116 1.69344 -9.08163
10.0977 10.9116 1.69344 -9.08163 -11.5071
10.9116 1.69344 -9.08163 -11.5071 -3.35299
1.69344 -9.08163 -11.5071 -3.35299 7.88384
-9.08163 -11.5071 -3.35299 7.88384 11.8723
```

Calculating the Determinant of a Matrix

Write C++ code to calculate the determinant of a matrix using the classes in uBLAS.

Solution

Calculating the determinant of a matrix is one of the classic problems in LA theory. Among other things, this value is used to determine if a system of equations (as expressed by the matrix of coefficients) has a unique solution.

To be able to easily compute the determinant of a matrix in C++, you can use some of the classes and functions contained in the boost uBLAS library. These functions make use of the matrix class, which is one of the uBLAS internal representations for matrices.

A common solution for this kind of problem uses a simple but elegant algorithm that is taught in any course of linear algebra. The general idea is to use a recursive strategy to calculate the determinant of small submatrices, until you find the determinant of the complete matrix. The algorithm used by the computeDeterminant function, however, is computationally more efficient because it uses the result of lower-upper (LU) decomposition. LU decompositions are a way to factor a matrix into lower and upper triangular components.

The function lu_factorize returns zero if the matrix is non-singular, which means that it can be inverted and its corresponding linear system solved using Gaussian elimination. The matrix is subsequently rearranged using the Gaussian elimination procedure. Additionally, a permutation matrix is used to record the steps of the elimination procedure.

Considering this information, the algorithm for determinant computation is encoded in the function computeDeterminant. It uses the values stored in the main diagonal and the information in the permutation matrix to compute the corresponding determinant for the given matrix. You can see the complete algorithm for this method in the next section.

Complete Code

Listing 7-3 shows you an example for the uBLAS libraries. The function determinantSample uses some of the templates in uBLAS to calculate the determinant of a matrix, as described in the previous section.

Listing 7-3. Determinant.cpp

```
//
// Determinant.cpp
#include <boost/numeric/ublas/matrix.hpp>
#include <boost/numeric/ublas/io.hpp>
#include <boost/numeric/ublas/lu.hpp>

namespace ublas = boost::numeric::ublas;
using std::cout;
using std::endl;
```

```
// The sign is calculated from a given permutation.
// Just flip the sign for each change in permutation.
int getDeterminantSign(const ublas::permutation_matrix<std::size_t>& pm)
{
    int sign = 1;
    for (int i = 0; i < pm.size(); ++i)
    {
        if (i != pm(i))
        {
            sign *= -1.0;
        }
    }
    return sign;
}

// returns the value of the determinant for matrix m
//
double computeDeterminant(ublas::matrix<double>& m)
{
    ublas::permutation_matrix<std::size_t> pm(m.size1());

    double det = 1.0;
    if (ublas::lu_factorize(m,pm))
    {
        det = 0.0;
    }
    else
    {
        for(int i = 0; i < m.size1(); i++)
        {
            det *= m(i,i);
        }
        det = det * getDeterminantSign(pm);
    }
    return det;
}
```

```
void determinantSample()
{
    ublas::matrix<double> M(3, 3);

    for (unsigned i = 0; i < M.size1() ; ++i)
    {
        for (unsigned j = 0; j < M.size2() ; ++j)
        {
            M(i,j) = sin(3 * j);
        }
    }

    double determinant = computeDeterminant(M);

    cout << " determinant value is " << determinant
         << " for matrix " << M << endl;
}
```

Conclusion

This chapter includes a few programming samples for linear algebra computation in C++. One of the goals in this presentation is to show how mathematically oriented code can be used by financial application programmers. Linear algebra is the basis for many of the computational techniques that will be explored in the next few chapters, such as mathematical programming and portfolio optimization.

In this chapter, I first introduced some of the important libraries for linear algebra. Since linear algebra is such a specialized area, the best approach for programmers is to use code that contains well-tested components written by experts in the field. The standard for computational mathematics in the area of basic linear algebra is the BLAS library. Although BLAS is a Fortran and C library, its concepts have been translated into many other languages. In this chapter, you have explored uBLAS, a component of the boost libraries that implements the same levels of functionality supported by BLAS. It does it, however, using modern C++ techniques such as classes and templates. This can be viewed as an easier way to achieve the functionality of BLAS while at the same time supporting a high-level C++ interface.

The first example in Listing 7-1 shows how to use uBLAS to implement basic (level 1) operations on vectors and scalars. The class VectorOperations shows how these basic concepts can be invoked using the uBLAS framework.

More advanced operations are available for matrices. The second example (Listing 7-2) contains information and code examples of how to interact with matrices and vectors in uBLAS. Simple operations that can be easily performed by uBLAS include scalar and vector multiplication of matrices, transposition, and matrix-matrix multiplication.

Listing 7-3's example shows how these concepts can be used together to calculate the determinant of a matrix. To facilitate the solution of this problem, you can use the LU factorization function provided by uBLAS. This shows how some of the sophisticated algorithms in these LA libraries can be easily used to solve practical problems.

In the next chapter, we will continue to explore mathematical tools used in financial applications. I will show you a few examples about interpolation, a technique that is frequently used to find trends in data sets, including financial data. Along with other computational techniques, interpolation is widely used in the development and analysis of trading strategies.

CHAPTER 8

Interpolation

Interpolation is a commonly used technique that approximates a mathematical function, based on a set of points given as input. Fast interpolation is the secret for high-performance algorithms in several areas of financial engineering, which will be explored in the next chapters. This chapter shows you a few programming examples that cover some of the most common aspects of interpolation techniques, along with their efficient implementation in C++. You will explore the main procedures used in applications and see examples of how they work in practice.

Here are some of the topics covered in this chapter:

- Interpolation examples: A brief discussion of examples that show the effectiveness of using interpolation in financial problems.

- Linear Interpolation: One of the simplest interpolation techniques, linear interpolation uses linear functions, which can be represented as line segments. The quick nature of this technique makes it one of the most used forms of interpolation for functions that are hard to compute.

- Polynomial interpolation: If smooth transitions between different parts of the function are required, then it is not possible to use a linear interpolation directly. Polynomial interpolation allows the use of a single function that approximates all of the given points through a high-degree polynomial. You will see how to construct this polynomial and return the desired function for each value of the domain.

© Carlos Oliveira 2021
C. Oliveira, *Practical C++20 Financial Programming*, https://doi.org/10.1007/978-1-4842-6834-6_8

Linear Interpolation

Write a solution, along with C++ code, for the problem of interpolating a given set of data points using linear approximations.

Solution

Interpolation is the process of finding a function (or set of functions) that can be used to approximate an unknown function, as determined by a given set of points. The input for this process is therefore the points you would like to interpolate. The result is a general function that can be used to compute the unknown values for points outside the input set.

For example, suppose that you are given a time series composed of a set of observations. It is frequently desirable to find a function that generated those points. This estimation of the unknown function may also be used to calculate the corresponding value for a required time instant.

Interpolation has been used in several areas of science and engineering as a way to approximate functions. This may be necessary either because the original function is truly unknown (e.g., in the case of empirical processes) or because such a function is very difficult to calculate exactly. In finance, interpolation also plays an important role, frequently as part of more complex algorithms. In the case of a financial time series, for example, interpolation allows practitioners to calculate values for a time series that are difficult to compute while using only an approximation for the unknown function. Moreover, in some of these applications, interpolation can also be used as a way to forecast values, at least for short periods of time. In this role, it can also be used as a simple forecasting component of trading algorithms.

In this section, you will see how to perform interpolation using linear functions. This is the simplest way to provide the interpolation of a set of points, since it requires only two points at a time in order to directly connect input values. For example, suppose that you're given a set of points, as displayed in Figure 8-1.

$y1 = (10,0.6)$, $y2 = (20,0.11)$, $y3 = (30,1.1)$, $y4 = (40,1.62)$, $y5 = (49,1.4)$.

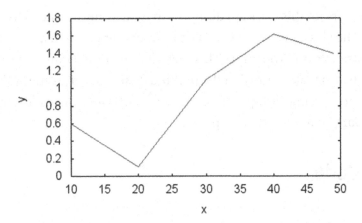

Figure 8-1. *Interpolation graph for points in the first example*

The points don't need to be evenly spaced, although that might work better when a linear interpolation is desired. Using these points, one can visualize a simple way to interpolate values. The strategy for creating an interpolation is to sort the input points based on their first dimension (the x axis on Figure 8-1) and join with a line the two points that are right before and after the desired value of x. The interpolation is then given by the formula in Step 4 of the list following Figure 8-1.

The algorithm for the linear interpolation of a set of points can, therefore, be summarized in the following steps:

1. Read the input values $(x[i],y[i])$ and the desired coordinate value x.

2. Sort the input in increasing order of the first coordinate $(x[i])$.

3. Calculate the first pair of points $(x[i],y[i])$, $(x[j],y[j])$ such that j=i+1 and x is within $x[i]$ and $x[j]$.

4. Use the following equation to determine the value of y corresponding to x:

$$y = \frac{x - x_1}{x_0 - x_1}$$

You can easily implement this algorithm in C++ so that the function is computed for each value of x. I present the LinearInterpolation as the main class responsible for storing the necessary data as well as calculating points using this interpolation technique.

237

The constructors for this class are designed to reduce any overhead for the creation of new objects. The class provides the `setPoints` member function as a way to define the known points of the interpolation. This member function retains the values passed as a parameter. It also makes sure that the points are stored in the order of increasing x values. This is done using a simple algorithm that sorts the x values (the standard function `std::sort` is used for this purpose).

Complete Code

```
//
//  LinearInterpolation.h

#ifndef __FinancialSamples__LinearInterpolation__
#define __FinancialSamples__LinearInterpolation__

#include <vector>

class LinearInterpolation {
public:
    LinearInterpolation();
    LinearInterpolation(const LinearInterpolation &p);
    ~LinearInterpolation();
    LinearInterpolation &operator=(const LinearInterpolation &p);
    void setPoints(const std::vector<double> &xpoints, const
    std::vector<double> &ypoints);
    double getValue(double x);
private:
    std::vector<double> m_x;
    std::vector<double> m_y;
};

#endif /* defined(__FinancialSamples__LinearInterpolation__) */

//
//  LinearInterpolation.cpp

#include "LinearInterpolation.h"
```

```cpp
LinearInterpolation::LinearInterpolation()
: m_x(),
  m_y()
{
}

LinearInterpolation::LinearInterpolation(const LinearInterpolation &p)
: m_x(p.m_x),
  m_y(p.m_y)
{
}

LinearInterpolation::~LinearInterpolation()
{
}

LinearInterpolation &LinearInterpolation::operator=(const
LinearInterpolation &p)
{
    if (this != &p)
    {
        m_x = p.m_x;
        m_y = p.m_y;
    }
    return *this;
}

void LinearInterpolation::setPoints(const std::vector<double> &xpoints,
                                    const std::vector<double> &ypoints)
{
    m_x = xpoints;
    m_y = ypoints;

    // update points to become sorted on x axis.
    std::sort(m_x.begin(), m_x.end());
```

```
    for (int i=0; i<m_x.size(); ++i)
    {
        for (int j=0; j<m_x.size(); ++j)
        {
            if (m_x[i] == xpoints[j])
            {
                m_y[i] = ypoints[j];
                break;
            }
        }
    }

}

double LinearInterpolation::getValue(double x)
{

    double x0=0, y0=0, x1=0, y1=0;
    if (x < m_x[0] || x > m_x[m_x.size()-1])
    {
        return 0; // outside of domain
    }

    for (int i=0; i<m_x.size(); ++i)
    {
        if (m_x[i] < x)
        {
            x0 = m_x[i];
            y0 = m_y[i];
        }
        else if (m_x[i] >= x)
        {
            x1 = m_x[i];
            y1 = m_y[i];
            break;
        }
    }
```

```
        return y0 * (x-x1)/(x0-x1) +  y1 * (x-x0)/(x1-x0);
}

int main()
{
    double xi = 0;
    double yi = 0;
    vector<double> xvals;
    vector<double> yvals;
    while (cin >> xi)
    {
        if (xi == -1)
        {
            break;
        }
        xvals.push_back(xi);
        cin >> yi;
        yvals.push_back(yi);
    }
    double x = 0;
    cin >> x;
    LinearInterpolation li;
    li.setPoints(xvals, yvals);
    double y = li.getValue(x);
    cout << "interpolation result for value " << x << " is " << y << endl;
    return 0;
}
```

Running the Code

For example, consider again the points in the example shown in Figure 8-1. To calculate a linear interpolation for value 27, you need to execute the application and enter the following data:

./linearInterpolation

```
10 0.6
20 0.11
30 1.1
40 1.62
49 1.4
-1
27
```

interpolation result for value 27 is 0.803

Polynomial Interpolation

Construct a polynomial interpolation for a given set of points in C++.

Solution

In the previous example, you saw how data points may be used to interpolate values in a continuous interval, through the use of piecewise linear equations. However, while it is possible to use linear interpolation in a large number of practical situations, the problem with this type of approximation is that the resulting curve is non-smooth. This means that it contains inflection points that mark transitions in the function, exactly at the interception of the different lines. In mathematical terms, it is said that such functions are non-differentiable because of this sudden transition. Such perceptible changes are undesired in some applications, and you may want to interpolate the values in such a way that the transition between observed points is seamless.

To avoid the described problems with the use of linear interpolation, a more sophisticated scheme may be employed, which uses higher-degree polynomials to smooth out the transitions. What is more important, a single polynomial found using this method can be used to interpolate all given data points at the same time. The result of this type of interpolation is that you just need a single polynomial equation to generate values for any desired input.

Polynomial interpolation is based on the mathematical fact that, given a polynomial with a high enough degree, you can find a corresponding polynomial function that passes through the exact points that are provided as input. This is guaranteed due to some well-known algebraic properties of polynomials. For example, suppose that we're given the sequence of points that follow (these are the same points used in the previous example for linear interpolation):

y1 = (10,0.6), y2 = (20,0.11), y3 = (30,1.1), y4 = (40,1.62), y5 = (49,1.4).

A polynomial interpolation algorithm would return a value based on a polynomial defined by a set of coefficients. Using that information, you can calculate any intermediate point or even points that are outside the given range of observations, since polynomials are typically defined for any real number. You can also use the calculated polynomial to plot the values of the interpolating function, as you see in Figure 8-2.

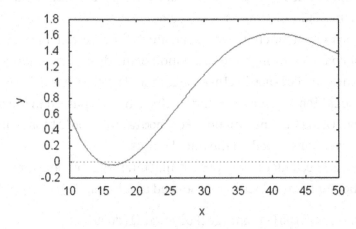

Figure 8-2. *A polynomial function to interpolate values using a small number of observations: notice how the polynomial function is smooth, unlike the solution in Figure 8-1, which uses line segments*

The technique used here to solve this polynomial interpolation problem is called Lagrange's interpolation algorithm. Using Lagrange's interpolation method, for each sequence of n+1 points (xi,yi), you can create a polynomial that has degree n and passes through these points. Using the points given as input, the general formula for the coefficients of the polynomial is given by

$$L_k(x) = \frac{(x - x_0)(x - x_1)\ldots(x - x_{k-1})(x - x_{k+1})\ldots(x - x_n)}{(x_k - x_0)(x_k - x_1)\ldots(x_k - x_{k-1})(x_k - x_{k+1})\ldots(x_k - x_n)}$$

Notice that this function skips the value of *xk* to avoid zero terms in the numerator and denominator. Now, the complete polynomial representation for interpolating the input values *xk* can be written using the coefficients *Lk(x)*.

$$P(x) = y_0 L_0(x) + \cdots + y_n L_n(x)$$

This is a function that can be used to provide the interpolation of any value, given the n+1 input observations (*xi, yi*). The proof for this formula is beyond our goals in this section, but notice that when the input value *x* is one of the known *xi*, then it will have a component *x – xi* that will result in zero in all cases, but for *Li(x)*. In that case, however, the numerator is the same as the denominator, which results in the value 1. Therefore, for these values, the solution is just *yi* as expected.

Using this polynomial function, we can create a C++ class that implements the interpolation mechanism through the simulation of the desired polynomial. This is achieved using the class `PolynomialInterpolation`. The class has a design similar to `LinearInterpolation`, storing the x and y values that are passed in the `setPoints` member function. Using that information, `PolynomialInterpolation` is able to perform the necessary calculations based on the initial points.

The real work is of calculating the polynomial interpolation performed in the getValue member function, which is reproduced as follows:

```
double PolynomialInterpolation::getPolynomial(double x)
{
    double polynomialValue = 0;

    for (size_t i=0; i<m_x.size(); ++i)
    {
        // compute the numerator
        double num = 1;
        for (size_t j=0; j<m_x.size(); ++j)
        {
            if (j!=i)
            {
                num *= x - m_x[j];
            }
        }
```

```
// compute the denominator
double den = 1;
for (size_t j=0; j<m_x.size(); ++j)
{
    if (j!=i)
    {
        den *= m_x[i] - m_x[j];
    }
}

// value for i-th term
polynomialValue += m_y[i] * (num/den);
}

return polynomialValue;
}
```

The calculation is done in an iterative way, where at each step of the for loop, one of the polynomials $Lk(x)$ is computed and added to the local variable polynomialValue. The internal part of the loop can be divided into three parts. In the first part, the numerator is calculated as a result of multiplying all of the terms $x - xj$, something that is not necessary when $i = j$. The second part is the calculation of the denominator, which is very similar to the first step, as you can confirm looking at the original formula. The values of the denominator are stored in the local variable den. The third step consists of multiplying the value of y by the fraction defined by the numerator and denominator that were computed in the previous two steps.

The complexity of the algorithm previously described is dependent on the given input. The more input values you provide, the longer this algorithm will take. The variation in time is quadratic with the number of input values, since for each value, we need to perform a for loop inside a second loop, and each such loop runs a number of times equal to the number of points used. In terms of computational complexity, this is said to have a time complexity of $O(n^2)$.

Complete Code

```
//
//
//  PolymonialInterpolation.h

#ifndef __FinancialSamples__PolymonialInterpolation__
#define __FinancialSamples__PolymonialInterpolation__

#include <vector>

class PolynomialInterpolation {
public:
    PolynomialInterpolation();
    PolynomialInterpolation(const PolynomialInterpolation &p);
    ~PolynomialInterpolation();
    PolynomialInterpolation &operator=(const PolynomialInterpolation &);
    void setPoints(const std::vector<double> &x, const std::vector<double> &y);
    double getPolynomial(double x);
private:
    std::vector<double> m_x;
    std::vector<double> m_y;
};

#endif /* defined(__FinancialSamples__PolymonialInterpolation__) */

//
//  PolymonialInterpolation.cpp

#include "PolymonialInterpolation.h"

PolynomialInterpolation::PolynomialInterpolation()
: m_x(),
  m_y()
{
}
```

```
PolynomialInterpolation::PolynomialInterpolation(const
PolynomialInterpolation &p)
: m_y(p.m_y),
  m_x(p.m_x)
{
}

PolynomialInterpolation::~PolynomialInterpolation()
{
}

PolynomialInterpolation &PolynomialInterpolation::operator=(const
PolynomialInterpolation &p)
{
    if (this != &p)
    {
        m_x = p.m_x;
        m_y = p.m_y;
    }
    return *this;
}

void PolynomialInterpolation::setPoints(const std::vector<double> &x,
                                        const std::vector<double> &y)
{
    m_x = x;
    m_y = y;
}

double PolynomialInterpolation::getPolynomial(double x)
{
    double polynomialValue = 0;

    for (size_t i=0; i<m_x.size(); ++i)
    {
        // compute the numerator
        double num = 1;
```

```
        for (size_t j=0; j<m_x.size(); ++j)
        {
            if (j!=i)
            {
                num *= x - m_x[j];
            }
        }

        // compute the denominator
        double den = 1;
        for (size_t j=0; j<m_x.size(); ++j)
        {
            if (j!=i)
            {
                den *= m_x[i] - m_x[j];
            }
        }

        // value for i-th term
        polynomialValue += m_y[i] * (num/den);
    }

    return polynomialValue;
}

int main()
{
    double xi = 0;
    double yi = 0;
    vector<double> xvals;
    vector<double> yvals;
    while (cin >> xi)
    {
        if (xi == -1)
        {
            break;
        }
```

```
        xvals.push_back(xi);
        cin >> yi;
        yvals.push_back(yi);
    }
    double x = 0;
    cin >> x;
    PolynomialInterpolation pi;
    pi.setPoints(xvals, yvals);
    double y = pi.getPolynomial(x);
    cout << "interpolation result for value " << x << " is " << y << endl;
    return 0;
}
```

Running the Code

To run the previous code, you need to first compile it using a C++ compiler such as gcc. Then you can execute the code as follows:

./polyInterpolation
```
10 0.6
20 0.11
30 1.1
40 1.62
49 1.4
-1
27
interpolation result for value 27 is 0.795433
```

The code was executed several times using the data just displayed. Figure 8-2 presents a plot of the resulting data. Notice that the plot shows a smooth function that passes through the set of points given as input. This demonstrates that the polynomial calculated by the presented code is a good, smooth interpolation for the given set of points.

Conclusion

In this chapter, you learned about interpolation, a mathematical technique used to find reasonable approximations for a function, given a data set of known values. Interpolation plays a role in financial data analysis since it provides a way to analyze and simplify the calculation of complicated functions. It may also allow one to perform a simple forecast of future price changes, as well as helping with a better understanding of past data. You have seen a few programming examples that illustrate the use of interpolation in the context of C++ programming.

Initially, you learned about linear interpolation, a simple method to interpolate values when just a few points of the original function are known. This technique uses only linear functions to perform the desired interpolation. You have seen an example C++ class that can be used to return interpolated results for any given value in the domain of the function.

Next, you saw how to interpolate function values using a better approximation technique provided by the application of polynomials. Using polynomials, you have the ability to create a smooth (differentiable) function that touches the given points in n+1 points, where n is the degree of the polynomial. You have learned a simple formula known as Lagrange's method, which can be used to create such polynomial interpolations from any given set of points. You have also learned about how to code a C++ class that implements this algorithm. I provided a complete example of how to use this class to generate a smooth interpolation of a given set of values.

In the next chapter, you will learn about another important mathematical skill used in financial applications. The calculation of roots of equations is a fundamental technique that can allow you to find solutions for many important problems in economics and engineering. You will see how the methods for finding roots of equations can be implemented and used in C++.

CHAPTER 9

Calculating Roots of Equations

Solving equations is one of the building blocks of many engineering and scientific algorithms. A typical example is the calculations needed for options and derivatives pricing, using the Black-Scholes model. As financial algorithms become more sophisticated, there is a great need to calculate the results of equations in general. These results are frequently used in the analysis of investments and in new trading strategies. It is important not only to be able to solve equations but also to calculate the roots of such equations in an efficient way.

In this chapter, you will learn some of the popular methods for calculating roots of equations. The coding samples presented here cover different methods of calculating equation roots, along with explanations of how they work and when they should be used. The following are some of the topics covered in this chapter:

- Bisection method: A simple method that explores the change in signal around the root of equations; this method is easy to implement. You will learn the basics of the bisection method and how to code it in C++.

- Secant method: The secant method is an improvement over the bisection method, which tries to use the value of the function in the given range to guide the position of the new approximation of the root. The secant method can in many cases speed up the search for a root.

- Newton's method: This method, sometimes also called the Newton/ Raphson method, uses the derivative of a function as the guide for the search of the root for a given equation. Since the derivative

251

© Carlos Oliveira 2021
C. Oliveira, *Practical C++20 Financial Programming*, https://doi.org/10.1007/978-1-4842-6834-6_9

determines the slope of the tangent to the function, its value can be used to calculate a new approximation. Successive values may converge to the desired solution, and an error parameter can be used to determine when to stop the process.

Bisection Method

Create a class implementing the bisection method to find roots of equations.

Solution

Finding the roots of an equation is the process of determining any points for which the corresponding function is zero. During the development of computational mathematics, several methods have been devised to calculate roots of equations. This chapter covers a few of the most common of these methods.

The bisection method tries to find the roots of equations by using a simple strategy: the idea is to look at the sign of the function at different points of the domain and use that information to decide if there is a root on that interval. For example, consider the function

$$f(x) = (x-1)^3$$

Figure 9-1 displays the graph of this function in the interval -1 to 3.

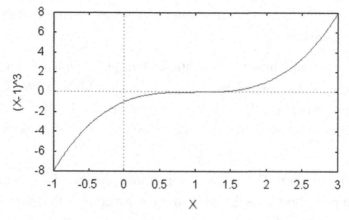

Figure 9-1. *Function $(x-1)^3$ in the interval -1 to 3*

This is a function that has a root for the value x=1 and has two areas that have distinct signs: for values less than x=1, the sign is negative. For values greater than 1, the sign of the function is positive. If you want to determine the exact place where the function is equal to zero, you can start from an interval where the sign of the function is different (in this case, you could use the interval -1 to 3) and look for the exact value where the function changes sign: that must be the location of one of the roots of the function.

Note   This argument works only when the function that you're dealing with is continuous. That is, there is no point where the function suddenly jumps from one point to another, which would make the foregoing argument invalid. Continuous functions are differentiable in the range where they're defined, so that is a way to know if a function is continuous. Most functions in economics, physics, and engineering have this property, so we assume that this is the case when the bisection method is applied.

With this intuitive insight, the bisection method tries to employ a systematic approach to determine the range tested and the location where the change of sign occurs. Essentially, the method is to bisect the original range and determine if the subrange still has different signs.

For example, using the same function, consider that we take the original range between -1 and 2. The middle of this range is ½, and therefore the algorithm will check the sign of $f\left(\frac{1}{2}\right) = \left(\frac{1}{2}-1\right)^3 = \left(-\frac{1}{2}\right)^3 = -\frac{1}{2^3} = -\frac{1}{8}$. Now, consider the sign of the two ending points of the initial range:

- $f(-1) = (-1-1)^3 = -8$, therefore with a negative sign.

- $f(2) = (2-1)^3 = 1^3 = 1$, therefore with a positive sign.

This means that between the values $x = -1$ and $x = \frac{1}{2}$, the sign is identical. On the other hand, between $x = \frac{1}{2}$ and $x = 2$, the sign changes. Therefore, the root must be in the place where the sign changes, which is somewhere between ½ and 2.

Another iteration of this process will calculate the value of $f\left(\frac{5}{4}\right) = \left(\frac{5}{4}-1\right)^3 = \left(\frac{1}{4}\right)^3 = 1/4^3$, which has a positive sign. Since $f(1/2)$ has a negative sign and $f(2)$ has a positive sign, the change in sign must happen in the interval from $\frac{1}{2}$ to $\frac{5}{4}$. This is easily observed to be true by checking the graph in Figure 9-1.

Notice that this process will systematically reduce the size of the range where we're searching for a root of the equation. At each step, we're decreasing the range by a half and getting closer to the location of the root. After a number of iterations, you will get a value that is as close as needed from the true root. This algorithm therefore can stop whenever the size of the remaining range is less than the desired error. For example, if the range is less than 0.001, and your desired error is 0.01, then stop.

Using the process we just described, we can now describe the algorithm for bisection in the following way:

1. Define an initial range (a,b) where you want to search for one or more roots of the equation.

2. Calculate the values of $f(a)$ and $f(b)$.

3. Determine the bisection point $c = \frac{a+b}{2}$ for the interval (a,b) and its corresponding function value $f(c)$.

4. If the signs of $f(a)$ and $f(c)$ are different, (a,c) becomes the new range for the algorithm. Otherwise, the new range is defined as (c,b).

5. If the new range has length less than the error (threshold) E, stop and report c as the solution. Otherwise, continue with step 1.

The preceding algorithm works in an iterative way, where the size of the range (a,b) is constantly reduced by half. Therefore, the number of steps it will perform is bounded by $\log_2 \frac{(b-a)}{E}$. There is a trade-off between number of iterations and desired approximation, which means that if you want a very small error, more iterations will be necessary.

The procedure described has been coded as the part of the BisectionSolver class, which can be used to find roots of equations. The interesting characteristic of BisectionSolver is that it is a generic solver for any continuous function. You just need to pass the function as a parameter to the constructor of the class.

For this purpose, I created a class called MathFunction that will be useful not only here but also for other classes that use functions as parameters. The MathFunction class defines what is called a function object, that is, an object that behaves like a function. This is important for our algorithms, because it allows one to use the object as if it were a function.

To create a functional object, you need a class that implements `operator()`, that is, the operator for function invocation. Moreover, I have defined the class `MathFunction` as a template class, so that you can pass the right return type when creating the concrete implementation. For example, you may want to define a `MathFunction` subclass that is defined for `float` values only. Or, conversely, you may have a function whose domain is the set of integers.

Finally, I want `MathFunction` to be just a root class, so that only concrete implementations can be instantiated. To make this possible, `MathFunction` is an abstract class, which is defined by using the `=0` notation after the declaration of `operator()`. Concrete subclasses of this class need to provide a concrete implementation of this operator. The example code shows how this can be done.

```
class F1 : public MathFunction<double> {
public:
        virtual ~F1() {}
        virtual double operator()(double value);
};

double F1::operator ()(double x)
{
        return (x-1)*(x-1)*(x-1);
}
```

This is a definition for the example function $f(x) = (x-1)\text{^}3$.

The implementation of the function `getRoot` is straightforward. The function starts with the interval given as a parameter and halves it using the criteria defined by the bisection method.

Complete Code

```
//
// MathFunction.h

#ifndef MATHFUNCTION_H_
#define MATHFUNCTION_H_
```

```cpp
template <class Res>
class MathFunction {
public:
        MathFunction();
        virtual ~MathFunction();
        virtual Res operator()(Res value) = 0;
};

#endif /* MATHFUNCTION_H_ */

//
// BisectionMethod.h

#ifndef BISECTIONMETHOD_H_
#define BISECTIONMETHOD_H_

template <class T>
class MathFunction;

class BisectionMethod {
public:
        BisectionMethod(MathFunction<double> &f);
        BisectionMethod(const BisectionMethod &p);
        ~BisectionMethod();
        BisectionMethod &operator=(const BisectionMethod &p);
        double getRoot(double x1, double x2);

private:
        MathFunction<double> &m_f;
        double m_error;
};

#endif /* BISECTIONMETHOD_H_ */

//
// BisectionMethod.cpp

#include "BisectionMethod.h"
```

```
#include "MathFunction.h"
#include <iostream>

using std::cout;
using std::endl;

namespace {
const double DEFAULT_ERROR = 0.001;
}

BisectionMethod::BisectionMethod(MathFunction<double> &f)
: m_f(f),
  m_error(DEFAULT_ERROR)
{
}

BisectionMethod::BisectionMethod(const BisectionMethod &p)
: m_f(p.m_f),
  m_error(p.m_error)
{
}

BisectionMethod::~BisectionMethod()
{
}

BisectionMethod &BisectionMethod::operator =(const BisectionMethod &p)
{
        if (this != &p)
        {
                m_f = p.m_f;
                m_error = p.m_error;
        }
        return *this;
}
```

```cpp
double BisectionMethod::getRoot(double x1, double x2)
{
        double root = 0;
        while (std::abs(x1 - x2) > m_error)
        {
                double x3 = (x1 + x2) / 2;
                root = x3;
                if (m_f(x1) * m_f(x3) < 0)
                {
                        x2 = x3;
                }
                else
                {
                        x1 = x3;
                }
                if (m_f(x1) * m_f(x2) > 0)
                {
                        cout << " function does not converge "  << endl;
                        break;
                }
        }
        return root;
}

 // ---- this is the implementation for an example function

namespace {

class F1 : public MathFunction<double> {
public:
        virtual ~F1();
        virtual double operator()(double value);
};

F1::~F1()
{
}
```

```cpp
double F1::operator ()(double x)
{
        return (x-1)*(x-1)*(x-1);
}

}

int main()
{
        F1 f;
        BisectionMethod bm(f);
        cout << " the root of the function is " << bm.getRoot(-100, 100) <<
        endl;
        return 0;
}
```

Running the Code

To run the code, along with the example provided, use a compiler such as gcc to generate the executable bisectionMethod. Then, you can run it to get results as follows:

./bisectionMethod
```
root is 0
root is 50
root is 25
root is 12.5
root is 6.25
root is 3.125
root is 1.5625
root is 0.78125
root is 1.17188
root is 0.976562
root is 1.07422
root is 1.02539
root is 1.00098
root is 0.98877
root is 0.994873
```

```
root is 0.997925
root is 0.999451
root is 1.00021
the root of the function is 1.00021
```

This shows the result of executing the bisection method on the function $f(x) = (x-1)\wedge3$, starting with the interval -100 to 100. I showed the intermediate steps just for clarity.

The Secant Method

Create a class to solve equations using the secant method.

Solution

In the previous programming example, you learned about the bisection method for the solution of equations. You can find the roots of an equation through the decomposition of the domain into a set of ranges, each of which you can test for changes in sign. If the sign of the function is different from the sign at the end of the interval, then it is possible to find a point where this function changes from positive to negative, making it the root of the equation.

While the bisection method can find solutions to a large number of equations, it is not the fastest method for this purpose. One of the reasons is that it doesn't use any of the properties of the function other than the sign at the extremes of its range. On the other hand, using additional information about the function values, for example, it would be possible, at least in principle, to achieve a faster convergence to the root of the function.

One of the ways to use the value of the function is contained in the algorithm called the secant method. The general idea of the secant method is to use the function value at the extremes of each interval as a way to approximate how close one is from the true root of the equation. In this way, it is possible to get closer to the root and reduce the number of iterations necessary to find the desired solution.

The secant of a function is the name given to the line connecting to points defined by that function. For example, given the function $f(x) = x^2$ in the range 1 to 4, the secant to that function in the given range is the line segment connecting the points (1,1) and (4,16), since these are the two points defined by function. We can generalize this concept to any function that is continuous in a particular range.

The secant method uses information derived from the secant to the function to define the new segment to be explored. To do this, the method calculates the point of intersection of the secant with the x-axis and uses that point to define the new segment. This is possible whenever the sign of the points at the beginning and end of a range is different, because in that case, the secant will intercept with the x-axis. As you may remember from the previous section, this is similar to the criterion used by the bisection method, with the difference that bisection uses the midpoint, instead of a point based on the secant of the function.

As an example of how this works in practice, consider the same function used in the section "Bisection Method": $f(x) = (x-1)^3$, in the range -1 to 2. In this case, we can calculate the values of $f(-1) = -8$ and $f(2) = 1$, which have different signs. We can, based on that information, use the secant to the function on this interval to find an intersection with the x-axis. The line segment we want to use is, therefore, connecting the points $(-1,-8)$ and $(2,1)$.

With a little of algebra, you will find that the slope of this line is

$$\frac{y_1 - y_0}{x_1 - x_0} = \frac{1-(-8)}{2-(-1)} = \frac{9}{3} = 3$$

And, since it is known that the point $(x_1,y_1) = (2,1)$ is touched, the secant line is given by

$$h(x) = y_1 + 3(x-x_1) = 1 + 3(x-2)$$

Now, the intersection point with the x-axis can be calculated using this equation and the fact that $h(x) = 0$ at the intersection (see Figure 9-2):

$$0 = 1+3x-6 = -5+3x$$

This means that $x = \frac{5}{3}$ is the intersection point between the secant and the x-axis. You can easily see, as shown in Figure 9-2, that the intersection of the secant with the x-axis is a point (let's call it x_2) that is one step closer to the root of the equation. This process can be repeated with the new interval defined by x_0 and x_2, until the root of the equation is approximated within a small error (which can be predefined by the algorithm).

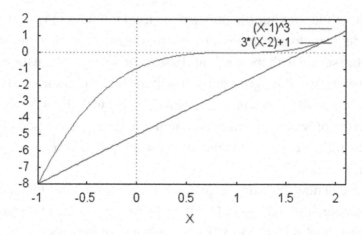

Figure 9-2. *The original function (x–1)³ and its secant in the interval -1 to 2*

Generalizing the $h(x)=$ equation shown just a bit earlier, the equation for the secant can be denoted as

$$h(x) = y_1 + \frac{y_1 - y_0}{x_1 - x_0}(x - x_1).$$

When this equation intercepts with the line y=0, we have

$$y_1 + \frac{y_1 - y_0}{x_1 - x_0}(x - x_1) = 0,$$

which yields the result

$$x = x_1 - \frac{y_1(x_1 - x_0)}{y_1 - y_0}.$$

With this equation, we can calculate the new point x that will be used at the next step of the algorithm. Summarizing the steps described, the secant method for finding roots of equations can be described in the following way:

1. Define an initial range (a,b) where you want to search for one or more roots of the equation.

2. Calculate the values of $f(a)$ and $f(b)$.

3. Determine the secant line using the equation

$$h(x) = f(b) + \frac{f(b) - f(a)}{b - a}(x - b)$$

4. Using this equation, find the intersection point

$$c = b - \frac{f(b)(b - a)}{f(b) - f(a)}$$

5. If the difference $|c - b|$ has length less than the error (threshold) E, stop and report c as the solution. Otherwise, continue with step 1.

It has been observed that for some functions, this algorithm converges to a solution more quickly than the bisection algorithm. This happens because the secant uses information that is already available with the function, which happens to make the intermediate point closer to the real solution.

You can find an implementation of the algorithm discussed in the SecantSolver class. The design of this class is similar to BisectionSolver, since the problem discussed is the same. The main change is the use of a different middle point selection procedure, which makes this algorithm a little different from that presented in BisectionSolver.

Complete Code

```
// SecantMethod.h
//

#ifndef SECANTMETHOD_H_
#define SECANTMETHOD_H_

template <class T>
class MathFunction;

class SecantMethod {
public:
        SecantMethod(MathFunction<double> &f);
        SecantMethod(const SecantMethod &p);
        SecantMethod &operator=(const SecantMethod &p);
        ~SecantMethod();
```

```cpp
        double getRoot(double x1, double x2);
private:
        MathFunction<double> &m_f;
        double m_error;
};

#endif /* SECANTMETHOD_H_ */
// SecantMethod.cpp
//

#include "SecantMethod.h"

#include "MathFunction.h"
#include <iostream>

using std::cout;
using std::endl;

namespace {
const double DEFAULT_ERROR = 0.001;
}

SecantMethod::SecantMethod(MathFunction<double> &f)
: m_f(f),
  m_error(DEFAULT_ERROR)
{
}

SecantMethod::SecantMethod(const SecantMethod &p)
: m_f(p.m_f),
  m_error(p.m_error)
{
}

SecantMethod::~SecantMethod()
{
}
```

```cpp
SecantMethod &SecantMethod::operator=(const SecantMethod &p)
{
        if (this != &p)
        {
                m_f = p.m_f;
                m_error = p.m_error;
        }
        return *this;
}

double SecantMethod::getRoot(double x1, double x2)
{
        double root = 0;
    double fa = m_f(a);
    double fb = m_f(b);
    double c = 0, fc = 0;
        do
        {
        c = b - fb*(b-a)/(fb-fa);
                root = c;
        fc = m_f(c);

        cout << "-> " << c << " " << fc << " " << endl; // this line just
        for demonstration

        a  = b;
        fa = fb;
        b  = c;
        fb = fc;
    }
    while (std::abs(a - b) > m_error);
        return root;
}
// ---- this is the implementation for an example function

namespace {
```

```
class F2 : public MathFunction<double> {
public:
        virtual ~F2();
        virtual double operator()(double value);
};

F2::~F2()
{
}

double F2::operator ()(double x)
{
        return (x-1)*(x-1)*(x-1);
}

}

int main()
{
        F2 f;
        SecantMethod sm(f);
        double root = sm.getRoot(-10, 10);
        cout << " the root of the function is " << root << endl;
        return 0;
}
```

Running the Code

After compiling the code presented previously, you can run it and get the following results, which show the solution for the sample equation $f(x) = (x-1)^3$:

```
./secantMethod
-> 2.92233 7.10369
-> 2.85268 6.35922
-> 2.25777 1.98976
-> 1.98685 0.96108
-> 1.73375 0.395035
-> 1.5571 0.172905
```

```
-> 1.41961 0.0738799
-> 1.31702 0.0318621
-> 1.23923 0.0136922
-> 1.18062 0.00589209
-> 1.13634 0.00253417
-> 1.10292 0.00109016
-> 1.07769 0.000468932
-> 1.05865 0.000201717
-> 1.04427 8.67704e-05
-> 1.03342 3.73252e-05
-> 1.02523 1.60558e-05
-> 1.01904 6.90655e-06
-> 1.01438 2.97092e-06
-> 1.01085 1.27797e-06
-> 1.00819 5.49731e-07
-> 1.00618 2.36472e-07
-> 1.00467 1.01721e-07
-> 1.00352 4.37562e-08
-> 1.00266 1.88221e-08
  the root of the function is 1.00266
```

Newton's Method

Create a C++ class to implement Newton's method for calculating roots of equations.

Solution

As you have seen in the last few sections, it is possible to find solutions for a large number of equations by just using a bisection method. You can also try to improve the rate of convergence using additional information from the function, in such a way that the secant of the function is used in the desired interval. Taking this idea one step further, you will arrive at one of the most used methods for solving equations, which is attributed to Isaac Newton.

Newton's method for finding roots of equations uses the derivative of the function as a first approximation to the location of the root. Similar to the previous two methods you have seen, the process is iterative, and at each step, you can get closer to the real root of the equation. At the end, you will have a solution that is within a very small error, which can be determined before the algorithm starts.

To understand how the method works, consider again the function $f(x) = (x-1)^3$, which we have been using as an example. The derivative of this function can be easily calculated, since this is a polynomial, and is given by $f'(x) = 3(x-1)^2$. Now, suppose that you start with an initial guess of what the root value might be (if there is no guess, consider a random value as the starting point). Call that initial guess x_0. At that point, we can calculate two values, $f(x_0)$ and $f'(x_0)$.

In Figure 9-3, you will find a plot of the function and its tangent at point $x_0 = 0$ for the example function given previously. At that point, the value of $f(x_0) = f(0) = -1$, and the value of $f'(x_0) = f'(0) = 3$. Since the derivative of the function at a particular point represents the slope of the tangent to the curve at that point, we can calculate a point x_1, which is determined by the line that is tangent with the given equation using the following formula:

$$x_1 = x_0 - \frac{f(x_0)}{f'(x_0)} = 0 - \frac{-1}{3} = \frac{1}{3}$$

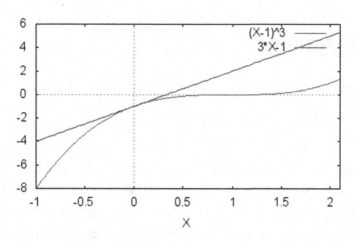

Figure 9-3. *The example function $f(x) = (x-1)^3$ and its tangent at point $x_0 = 0$. You may notice that the tangent intersects the x-axis at point $x_1 = \frac{1}{3}$. This is the first step in finding the root for the given function using Newton's method*

Once the intersection of the x-axis and the tangent has been found, you have a new starting point for the determination of the root for the given equation. Notice that, each time a new point is found, the algorithm gets a little closer to the desired point, although it may take a few iterations to achieve the desired precision. As in the previous two cases, you can determine the precision as a parameter to the algorithm and stop the computation once the difference between two successive approximations is less than the given parameter. This shows that the solution is converging to a point where the root is located.

In summary, Newton's method works by successively finding points that are determined by the tangent to the function. As you get closer to the root, the difference between the intersection of the tangent with the x-axis will get smaller. The stopping condition is that difference between the last and the current values is less than a given parameter.

Based on the previous information, the algorithm for Newton's method can be given as follows:

1. Define an initial value x, from which you want to search for one or more roots of the equation.

2. Given the input function $f(x)$, determine the derivative of function $f'(x)$.

3. Calculate the values of $f(x)$ and $f'(x)$ for the desired value.

4. Using the value $f'(x)$ as the slope of the tangent at the point x, calculate a new point x_1 using the following equation:

$$x_1 = x - \frac{f(x)}{f'(x)}.$$

5. Calculate the difference between x and x_1 as $e = |x - x_1|$.

6. If the value of e is less than the input error (threshold) E, stop and report x as the solution. Otherwise, rename x_1 to x and continue with step 1.

The fact that Newton's method depends not only on the function but also on its derivative can be seen as an advantage as well as a disadvantage. Sometimes, it is very easy to compute the derivative of a function, such as for polynomial and common trigonometric functions, but that is not always the case. However, the greatest advantage of Newton's method is that, unlike the bisection and secant methods, it works even when there is no difference in function sign.

Consider, for example, the function $f(x) = (x-1)^2$. Its graph is shown in Figure 9-4. Notice that, unlike other functions that you saw in this chapter, there is no place where it changes sign. Therefore, without some changes to the bisection method, it is not possible to find a root in this case. On the other hand, the function clearly has a derivative, being differentiable everywhere, and this makes it possible to find the root using Newton's method.

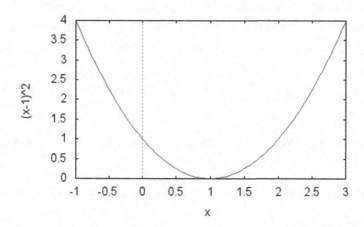

Figure 9-4. *A continuous, quadratic function $(x-1)^2$ that never changes sign but has a single root at x = 1*

A small issue that we need to consider when using Newton's method is the possibility that the derivative is zero at some point. If that is the case, then the next point is undefined, since a division by zero is not permitted. This difficulty can be avoided, however, if the algorithm adds a small value to the current point as a way to avoid the issue. More sophisticated techniques to solve this problem exist, however, as the reader will be able to find in one of the many existing books on the topic of numerical algorithms.

The previous algorithm was coded in C++ in the class NewtonMethod, which provides the necessary support for all the steps described. The design of the class is very similar to what you saw for the bisection method. Unlike the bisection method and the secant method, however, NewtonMethod depends not only on the function but also on its derivative. That's why in the example code, you will see a reference to two functions: F3 and DF3. They are necessary so that the NewtonMethod class knows how to find the value of the function and its derivative.

Complete Code

```
//
// NewtonMethod.h

#ifndef NEWTONMETHOD_H_
#define NEWTONMETHOD_H_

template <typename T>
class MathFunction;

class NewtonMethod {
public:
        NewtonMethod(MathFunction<double> &f, MathFunction<double>
        &derivative);
        NewtonMethod(const NewtonMethod &p);
        virtual ~NewtonMethod();
        NewtonMethod &operator=(const NewtonMethod &p);

        double getRoot(double initialValue);
private:
        MathFunction<double> &m_f;
        MathFunction<double> &m_derivative;
        double m_error;
};

#endif /* NEWTONMETHOD_H_ */

//
// NewtonMethod.cpp

#include "NewtonMethod.h"

#include "MathFunction.h"

#include <iostream>
```

```cpp
using std::endl;
using std::cout;

namespace {
const double DEFAULT_ERROR = 0.001;
}

NewtonMethod::NewtonMethod(MathFunction<double> &f, MathFunction<double>
&derivative)
: m_f(f),
  m_derivative(derivative),
  m_error(DEFAULT_ERROR)
{
}

NewtonMethod::NewtonMethod(const NewtonMethod &p)
: m_f(p.m_f),
  m_derivative(p.m_derivative),
  m_error(p.m_error)
{
}

NewtonMethod::~NewtonMethod()
{
}

NewtonMethod &NewtonMethod::operator=(const NewtonMethod &p)
{
        if (this != &p)
        {
                m_f = p.m_f;
                m_derivative = p.m_derivative;
                m_error = p.m_error;
        }
        return *this;
}
```

```cpp
double NewtonMethod::getRoot(double x0)
{

    double x1 = x0;
    do
    {
        x0 = x1;
        cout << " x0 is " << x0 << endl;  // this line just for
                                          demonstration
        double d = m_derivative(x0);
        double y = m_f(x0);
        x1 = x0 - y / d;
    }
    while (std::abs(x0 - x1) > m_error);

    return x1;

}

// ---- this is the implementation for an example function and its
derivative

namespace {

class F3 : public MathFunction<double> {
public:
        virtual ~F3();
        virtual double operator()(double value);
};

F3::~F3()
{
}

double F3::operator ()(double x)
{
        return (x-1)*(x-1)*(x-1);
}
```

```
class DF3 : public MathFunction<double> {
public:
        virtual ~DF3();
        virtual double operator()(double value);
};

// represents the derivative of F3
DF3::~DF3()
{
}

double DF3::operator ()(double x)
{
        return 3*(x-1)*(x-1);
}

}

int main()
{
        F3 f;
        DF3 df;
        NewtonMethod nm(f, df);
        cout << " the root of the function is " << nm.getRoot(100) << endl;
        return 0;
}
```

Running the Code

The code can be compiled and linked using a compiler such as gcc on UNIX. To run the resulting program and see its associated results, use the following command line:

./newtonMethod
```
x0 is 100
 x0 is 67
 x0 is 45
 x0 is 30.3333
 x0 is 20.5556
```

```
x0 is 14.037
x0 is 9.69136
x0 is 6.79424
x0 is 4.86283
x0 is 3.57522
x0 is 2.71681
x0 is 2.14454
x0 is 1.76303
x0 is 1.50868
x0 is 1.33912
x0 is 1.22608
x0 is 1.15072
x0 is 1.10048
x0 is 1.06699
x0 is 1.04466
x0 is 1.02977
x0 is 1.01985
x0 is 1.01323
x0 is 1.00882
x0 is 1.00588
x0 is 1.00392
x0 is 1.00261
the root of the function is 1.00174
```

The algorithm executes the steps for Newton's method for the function $f(x) = (x-1)^3$. Notice that even when starting from a distant value of 100, the algorithm converged to the solution 1.0 within the required error.

Conclusion

In this chapter, you have seen a few examples that deal with the search for roots of equations. This is a topic that is frequently useful in the solutions of equations appearing in financial engineering. Algorithms for trading and for investment analysis frequently require the solution of equations, which makes it necessary to find efficient techniques for the determination of equation roots.

In this chapter, I presented some of the most common techniques for the solution of equations that appear in finance. More specialized algorithms exist, however, and you can use the numerical methods literature as a starting point to explore the most recent methods.

The programming samples in this chapter show, initially, how to compute the root solution for equations using the bisection method. With the bisection algorithm, the desired range of the domain is divided evenly, and at each step, possible location of the root is approximated with higher accuracy. Due to the sign of the function in each end point of the range, it is possible to detect if a root of the equation is contained in that range. At the end of the procedure, one can determine within a small margin of error the location where the equation becomes zero.

Next, you have seen a programming sample for the secant method to solve equations. The method is based on the use of the secant line to the function in the interval that is being considered. When the secant intersects the x-axis, the new point is usually closer to the root of the equation. Performing several iterations of this procedure, it is possible to find the desired solution in less time than needed by the bisection method.

You have also learned about the most popular method for the determination of equation root, known as Newton's method. With this algorithm, a solution can be found through the use of the tangent to the function at a given point. Since the derivative of a function gives the slope of the function at each point, it is possible to use the derivative to find a reasonable approximation. As the algorithm iterates through successive points, the approximation gets better. As with other methods, the algorithm is generally stopped when a preset maximum error is achieved.

In the next chapter, I will talk about another important computational tool that is heavily used in financial algorithms: numerical integration. With algorithms for numerical integration, it is possible to find solutions for several difficult problems with a high degree of precision. You will also see how to implement these techniques using C++.

CHAPTER 10

Numerical Integration

Integrating a function is a common step in many financial algorithms. For example, some financial techniques that involve the use of differential equations depend on the evaluation of complex integrals. Such areas include derivatives pricing, insurance, and related algorithms.

However, when using these methods, you can find integrals that have no known analytical solution and need to be integrated numerically. Even if an equation can be integrated analytically, it may be more efficient to perform this task using numerical algorithms. For this purpose, this chapter explores some of the common ways of performing numerical integration. After reading this and the next chapters, you will have a better understanding of how these numerical integration algorithms work in practice and how to use them in your own projects.

We discuss programming examples that can readily be applied in the use of some common integration methods. We also discuss their performance and the accuracy of such numerical methods when implemented using the C++ language. The programming examples in this chapter cover the following topics:

- Midpoint method: A simple method of integration that uses an easy-to-compute approximation based on the midpoint of each integration interval.

- Trapezoid method: A more accurate method of numerical integration that employs a trapezoidal approximation to the integrated area.

- Simpson's method: A popular technique of numerical integration, Simpson's method provides a slightly better approximation than the previous two methods.

- Graphical examples of these solution methods: We present a graphical explanation of how these methods work, along with the code necessary to implement them.

© Carlos Oliveira 2021
C. Oliveira, *Practical C++20 Financial Programming*, https://doi.org/10.1007/978-1-4842-6834-6_10

The Midpoint Method

In this section, we create a C++ class to integrate functions using the midpoint method.

Solution

Integrating a function means, in a few words, finding the area of the curve formed by the function and the x-axis, when considering a single dimension. While the concept is simple, there is a large amount of literature concerning the practical importance of this problem. The most important result, also known as the fundamental theorem of calculus, is that integration is the inverse function of the derivative. In other words, applying the derivative to the integral of a function will lead to the original function. You can also integrate the derivative of a function to reach the original (up to a constant).

Finding the integral of a function by algebraic means is highly dependent on the previous definition. This means that you need to know a second function whose derivative is the function you want to integrate. In that case, the second function is the integral using the fundamental theorem of calculus. The problem is that it is not always possible to find an antiderivative using such methods. This leads to the need to determine the integral using the computational method.

There are several methods that can be used to integrate a function, but they all include the strategy of dividing the area of the desired function into many subareas and adding them all. The good thing about numerical integration is that most schemes of subdividing the area as described previously are convergent. This means that the solution for most functions can be used by any of the methods we discuss. What make these methods different is the computational effort and possibly some better convergence for a particular function or application.

We start our discussion with an algorithm commonly known as the midpoint method. This method was so named because of the use of a midpoint approximation to the desired area. Consider the function $f(x)=x^2+1$, and try to calculate the integral for this function in the interval 1 to 5. You can see a plot of this function in Figure 10-1.

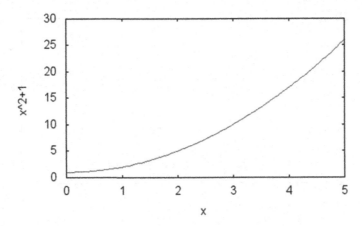

Figure 10-1. *Plot of function f(x)=x²+1 in the interval 1 to 5*

Of course, it would be easy to solve this problem using symbolic techniques, since there is a well-known way to determine the integral of a polynomial function. However, consider how the problem could be solved using a purely computational strategy.

The goal of the algorithm is to develop an approximation so that you can determine the considered integral within a prespecified error threshold. The first step is to devise a function that could approximate the given function in the given interval. It turns out that the easiest function we can try is the constant function $f(x)=c$.

Suppose that we use the constant function to approximate the integral in the interval 1 to 5. We can take as the value of the constant the average value of $f(x)$ in that interval, as determined by the extreme points.

That constant value would be $f\left(\dfrac{1+5}{2}\right) = f(3) = 10$. Check in Figure 10-2 how this constant function compares with the original function.

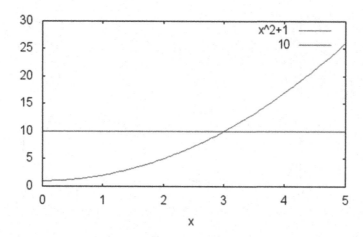

Figure 10-2. *The original function f(x) = x²+1 compared to the constant function* f(x) = c

The value of the approximation can be calculated as (5–1)10 = 40. You can compare this with the value defined by the closed calculation of the integral: the antiderivative of f(x) = x²+1 is $F(x) = \dfrac{x^3}{3} + x$, and using this equation as the definite integral on the interval 1 to 5, you will find the value 45.33. From this, you see that there is a large error between the correct value and the estimation using a single midpoint calculation.

The good news is that we can improve this approximation by considering smaller intervals over the required function and adding these values together. This is the basic technique you will see in this and in the next sections. Therefore, to improve the approximation in the preceding example, we can just divide the interval 1–5 into two intervals: 1 to 3 and 3 to 5.

Considering the average value for the constant function, you will find values $\left(\dfrac{1+3}{2}\right)^2 + 1 = 5$ for the first half of the interval and $\left(\dfrac{3+5}{2}\right)^2 + 1 = 17$ for the second half. This gives us an approximation of 2 × 5 + 2 × 17 = 44. Since the exact value is 45.33, you see that the approximation of 44 is much closer to the real value. You can see the new approximation in Figure 10-3.

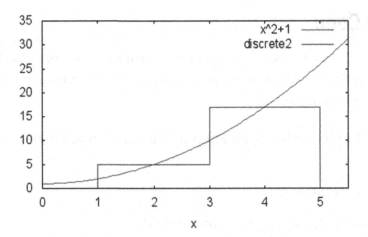

Figure 10-3. *Approximating the integral of f(x) = x²+1 using two constant values*

As you have seen, the secret of getting a great result from the midpoint method is subdividing the desired interval in smaller units and adding them, quite similar to the way the integral function is defined itself. While you could start with a single constant number and then divide the interval successively, it is much easier to start with a known large number of segments and expand the number of divisions if necessary. Using this strategy results in the following algorithm:

Define a range (A,B) where you want to calculate the integral to the equation.

1. Subdivide the initial range into N subintervals of equal sign.

2. Initialize the integral value S to zero.

3. For each subinterval (a,b), do the following:

    a. Take the values of a and b.

    b. Determine the middle point value:

$$m(a,b) = f\left(\frac{a+b}{2}\right).$$

    c. Add $m(a,b)$ to the integral.

The implementation of this simple method can be found in the MidpointIntegration class.

Complete Code

You can find the complete code that implements the method just described in Listing 10-1. The listing includes a header file for class `MidpointIntegration` as well as an implementation file.

Listing 10-1. Implementation for Midpoint Integration Method

```
//
// MidpointIntegration.h

#ifndef __FinancialSamples__MidpointIntegration__
#define __FinancialSamples__MidpointIntegration__

template <class T>
class MathFunction;

class MidpointIntegration {
public:
    MidpointIntegration(MathFunction<double> &f);
    MidpointIntegration(const MidpointIntegration &p);
    ~MidpointIntegration();
    MidpointIntegration &operator=(const MidpointIntegration &p);

    void setNumIntervals(int n);
    double getIntegral(double a, double b);

private:
    MathFunction<double> &m_f;
    int m_numIntervals;

};
#endif /* defined(__FinancialSamples__MidpointIntegration__) */

//
// MidpointIntegration.cpp

#include "MidpointIntegration.h"

#include "MathFunction.h"

#include <iostream>
```

```cpp
using std::cout;
using std::endl;

namespace  {
    const int DEFAULT_NUM_INTERVALS = 100;
}

MidpointIntegration::MidpointIntegration(MathFunction<double> &f)
: m_f(f),
  m_numIntervals(DEFAULT_NUM_INTERVALS)
{
}

MidpointIntegration::MidpointIntegration (const MidpointIntegration &p)
: m_f(p.m_f),
  m_numIntervals(p.m_numIntervals)
{
}

MidpointIntegration::~MidpointIntegration()
{
}

MidpointIntegration &MidpointIntegration::operator=(const
MidpointIntegration &p)
{
    if (this != &p)
    {
        m_f = p.m_f;
        m_numIntervals = p.m_numIntervals;
    }
    return *this;
}

void MidpointIntegration::setNumIntervals(int n)
{
    m_numIntervals = n;
}
```

```cpp
double MidpointIntegration::getIntegral(double a, double b)
{
    double S = 0;
    double intSize = (b - a)/m_numIntervals;
    double x = a;

    for (int i=0; i<m_numIntervals; ++i)
    {
        double midpt = m_f(x+(intSize/2));
        S += intSize * midpt;
        x += intSize;
    }
    return S;
}

// Example function  x^2 + 1

namespace  {

class F1 : public MathFunction<double>
{
public:
    ~F1();
    double operator()(double x);
};

F1::~F1()
{
}

double F1::operator ()(double x)
{
    return x*x+1;
}

}
```

```
int main()
{
    F1 f;
    MidpointIntegration mpi(f);
    double integral = mpi.getIntegral(1, 5);
    cout << " the integral of the function is " << integral << endl;

    mpi.setNumIntervals(200);
    integral = mpi.getIntegral(1, 5);
    cout << " the integral of the function with 200 intervals is " <<
    integral << endl;
    return 0;
}.
```

Running the Code

You can generate a binary executable from the source code in Listing 10-1 using any standards-compliant compiler such as gcc. Then, you can execute the code to get sample results such as the following, for the sample equation $f(x) = x^2 + 1$:

```
./midpointIntegration
the integral of the function is 45.3344
the integral of the function with 200 intervals is 45.3336
```

Notice that the solution tests the approximation for two cases: when the number of intervals is 100 (the default) and when the number of intervals is 200. Since the exact value of the function is $45 + \dfrac{1}{3}$, this shows an improvement in the result with an error going from the third decimal place to the fourth decimal place. You can improve the approximation for different functions or required errors by increasing the number of intervals if necessary.

Trapezoid Method

In this section, we create a C++ class that implements the trapezoid method for definite integral calculation.

Solution

As you have seen in the last section, it is not difficult to come up with an approximation for the integral of a function. However, in many applications, it is useful to have a faster and more efficient way to determine the definite integral of a function. This is especially true when the function that needs to be integrated is difficult to compute in the first place. In those situations, it is better to use an approximation technique that might be able to provide more accurate solutions to the integration problem.

In this coding example, I examine an alternative way to calculate the integral of a continuous function, called the trapezoid method. As the name indicates, the trapezoid method uses a geometric, intuitive idea to render the value under the curve for a particular function, in such a way that the resulting approximation is closer to the desired value.

To use the trapezoid method, we look at the integration problem using a geometric intuition about the best way to approximate the desired curve. Consider the function $f(x) = \sin(x)$ in the range $\dfrac{1}{2}$ to $\dfrac{5}{2}$. The desired integral is defined as the area under the curve. A simple approach to approximate this value is to use the area of linear function that approximates $\sin x$ between the extremes of the given interval. Using a graphical approach, you can see the results in Figure 10-4.

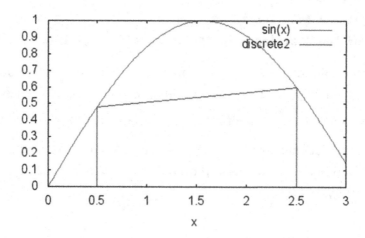

Figure 10-4. *Approximating the integral of f(x)=sin(x) over the interval from 1/2 to 5/2 with the help of one iteration of the trapezoid method*

The trapezoid method applied to that interval gives a relatively poor approximation. The real value of the indicated integral, when computed using symbolic techniques, is $\cos\dfrac{1}{2}-\cos\dfrac{5}{2}$, which is approximately 1.6787. The trapezoid method, on the other hand, yields the value $2\cdot\sin\dfrac{1}{2}+\sin\dfrac{5}{2}-\sin\dfrac{1}{2}=\sin\dfrac{1}{2}+\sin\dfrac{5}{2}\approx1.0778$.

Although this is a poor approximation, you can do better if you divide the interval of the desired function in two or more subareas. As you do this, the errors will become smaller, and the resulting value will be closer to the real value of the integral. For example, I will show how to improve the approximation of the previous function using the two subintervals from 1/2 to 3/2 and from 3/2 to 5/2.

The values of $f\left(\dfrac{1}{2}\right)=\sin\dfrac{1}{2}\approx0.4794$ and $f\left(\dfrac{3}{2}\right)=\sin\dfrac{3}{2}\approx0.9974$ result in a trapezoid with an area of 0.7384. The second interval, on the other hand, has value determined by $f\left(\dfrac{3}{2}\right)=\sin\dfrac{3}{2}$ and $f\left(\dfrac{5}{2}\right)=\sin\dfrac{5}{2}\approx0.5984$. The resulting trapezoid has area equal to 1.5364, which is closer to the exact value of 1.6787. You can see how this approximation works graphically in Figure 10-5.

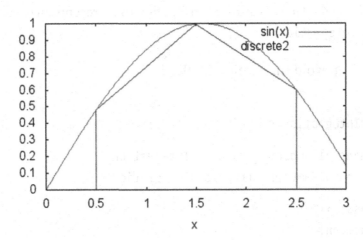

Figure 10-5. *Using the trapezoid method to approximate the area under the function f(x)=sin x, with two intervals (1/2 to 3/2 and 3/2 to 5/2)*

It can be proved, as shown in the previous two examples, that as the number of subintervals is increased, the quality of the approximation gets better. As a result, you can get as close as desired from the true value of the definite integral by increasing the number of subintervals in the trapezoid method.

I present a class, called `TrapezoidIntegration`, which shows how to implement the trapezoid method for any function passed as an argument. The implementation is made generic with the use of the `MathFunction` class. Passing a new object of the desired `MathFunction` class, you can calculate definite integrals for different functions using the `getIntegral` member function.

With the `TrapezoidIntegration` class, you can also determine the desired number of intermediate intervals used, if you use the member function `setNumIntervals`. This, as a result, allows you to reduce the error in the estimates of the definite integral, if necessary. Another thing you can do using the `setNumIntervals` is to reduce the computational effort necessary, by reducing the number of iterations of the algorithm. In this way, you have complete control over the trade-off between degree of approximation and computational efficiency.

Complete Code

Listing 10-2 is a complete implementation for the trapezoid method for integration as discussed in the previous section. You will find this code divided into a header file and an implementation file. There is also a `main` function that presents an example for the class `TrapezoidIntegration`.

Listing 10-2. Trapezoid Integration Method

```
//
//  TrapezoidIntegration.h

#ifndef __FinancialSamples__TrapezoidIntegration__
#define __FinancialSamples__TrapezoidIntegration__

template <class T>
class MathFunction;

class TrapezoidIntegration {
public:
    TrapezoidIntegration(MathFunction<double> &f);
    TrapezoidIntegration(const TrapezoidIntegration &p);
    ~TrapezoidIntegration();
    TrapezoidIntegration &operator=(const TrapezoidIntegration &p);
```

```
    void setNumIntervals(int n);
    double getIntegral(double a, double b);

private:
    MathFunction<double> &m_f;
    int m_numIntervals;

};

#endif /* defined(__FinancialSamples__TrapezoidIntegration__) */

//
//  TrapezoidIntegration.cpp

#include "TrapezoidIntegration.h"

#include "MathFunction.h"

#include <iostream>
#include <cmath>

using std::cout;
using std::endl;

namespace  {
    const int DEFAULT_NUM_INTERVALS = 100;
}

TrapezoidIntegration::TrapezoidIntegration(MathFunction<double> &f)
: m_f(f),
  m_numIntervals(DEFAULT_NUM_INTERVALS)
{
}

TrapezoidIntegration::TrapezoidIntegration (const TrapezoidIntegration &p)
: m_f(p.m_f),
  m_numIntervals(p.m_numIntervals)
{
}
```

```cpp
TrapezoidIntegration::~TrapezoidIntegration()
{
}

TrapezoidIntegration &TrapezoidIntegration::operator=(const
TrapezoidIntegration &p)
{
    if (this != &p)
    {
        m_f = p.m_f;
        m_numIntervals = p.m_numIntervals;
    }
    return *this;
}

void TrapezoidIntegration::setNumIntervals(int n)
{
    m_numIntervals = n;
}

double TrapezoidIntegration::getIntegral(double a, double b)
{
    double S = 0;
    double intSize = (b - a)/m_numIntervals;
    double x = a;

    for (int i=0; i<m_numIntervals; ++i)
    {
        double midpt = (m_f(x) + m_f(x+intSize))/ 2;
        S += intSize * midpt;
        x += intSize;
    }
    return S;
}

// Example function

namespace  {
```

```cpp
class F2 : public MathFunction<double>
{
public:
    ~F2();
    double operator()(double x);
};

F2::~F2()
{
}

double F2::operator()(double x)
{
    return sin(x);
}

}

int main()
{
    F2 f;
        TrapezoidIntegration ti(f);
    double integral = ti.getIntegral(0.5, 2.5);
        cout << " the integral of the function is " << integral << endl;

    ti.setNumIntervals(200);
    integral = ti.getIntegral(0.5, 2.5);
        cout << " the integral of the function with 200 intervals is " <<
        integral << endl;
    return 0;
}
```

Running the Code

You can compile the code presented in Listing 10-2 using a standards-compliant compiler such as gcc, Visual Studio, or llvm. After you compile the code, you can run the resulting application to test the results. The following is a sample of the program execution:

```
./trapezoidMethod
```
the integral of the function is 1.67867
the integral of the function with 200 intervals is 1.67871

The program displays the value of the integral of $\sin(x)$ from $1/2$ to $5/2$. The approximation is given for two different settings of the number of subintervals. The first result is for 100 subintervals. The second result shows the approximation achieved when that number of subintervals is doubled.

As in the previous example, it is possible to control the quality of the approximation by increasing the number of subintervals. Also, you can reduce that number in case you prefer to get quicker results.

Using Simpson's Method

Implement Simpson's method for definite integral calculation.

Solution

You have seen two common ways to calculate the value of the definite integral for a given continuous function. A third method, known as Simpson's method, is presented in this programming example. As with any technique for numeric integration, the general idea is to create a second function that approximates the desired function and apply it to several subintervals of the original domain until you get a good approximation.

Simpson's method, unlike the previous two methods that use linear approximations to the given function, employs a second-order polynomial to achieve a better convergence. In this way, instead of relying on a linear function to achieve the desired result, the approximation proposed with Simpson's method is better adapted to the behavior of the original curve.

The way Simpson's method work can be easily visualized with an example. Suppose you want to integrate the function used in the previous section: $f(x) = \sin x$. This function, being trigonometric, has no simple finite representation as a polynomial. However, it is possible to find very good approximations for a representation if you restrict the search to a small part of the function.

For example, I have shown in Figure 10-6 how it is possible to use a second-order polynomial function to approximate the value of $\sin x$ in the interval $1/2$ to $5/2$. Observe that the similarity between these two curves is good enough only in the short range of values inside the given interval, and outside that interval, these two functions vary widely.

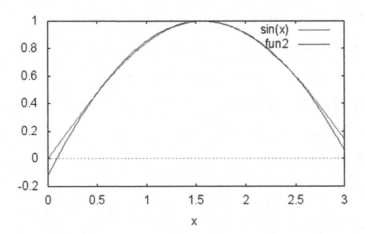

Figure 10-6. *Using a second-degree polynomial to approximate the value of f(x)=sin x in the interval 1/2 to 5/2*

The same idea is used in Simpson's method. Since a quadratic approximation may be so close to the desired function, the use of quadratic functions may dramatically improve the value of the definite integral calculated in this way. In fact, experiments have shown that Simpson's method has better accuracy than other algorithms, such as the midpoint method and the trapezoid method.

Note   The additional accuracy of Simpson's method can make it possible to reduce the number of subintervals necessary for the calculation of the definite integral. However, since you need to use a quadratic approximation instead of a simple linear function, the computational effort of each iteration will be higher. In the end, while Simpson's method produces superior results for most functions, users need to be aware of a possible trade-off in terms of computational time per iteration.

The second-order polynomial used in Simpson's method is defined in the following equation, which can be used to directly implement the proposed rule:

$$\frac{b-a}{6}\left[f(a)+4f\left(\frac{a+b}{2}\right)+f(b)\right]$$

Therefore, you can summarize the general algorithm as follows:

1. Define a range (A,B) where you want to calculate the integral to the equation.

2. Subdivide the initial range into N subintervals of equal sign.

3. Initialize the integral value S to zero.

4. For each subinterval (a,b), do the following:

   a. Take the values of a and b.

   b. Determine the approximation to the integral in the interval (a,b) given by the equation

   $$m(a,b) = \frac{b-a}{6}\left[f(a) + 4f\left(\frac{a+b}{2}\right) + f(b)\right]$$

   c. Add $m(a,b)$ to the integral.

This algorithm has been implemented as part of SimpsonsIntegration class

Complete Code

You can find the complete implementation of Simpson's method in Listing 10-3. The implementation presented there is later used in the main function.

Listing 10-3. Code for Simpson's Integration Method

```
//
// SimpsonsIntegration.h

#ifndef __FinancialSamples__SimpsonsIntegration__
#define __FinancialSamples__SimpsonsIntegration__

template <class T>
class MathFunction;

class SimpsonsIntegration {
public:
    SimpsonsIntegration(MathFunction<double> &f);
```

```cpp
    SimpsonsIntegration(const SimpsonsIntegration &p);
    ~SimpsonsIntegration();
    SimpsonsIntegration &operator=(const SimpsonsIntegration &p);

    double getIntegral(double a, double b);
    void setNumIntervals(int n);
private:
    MathFunction<double> &m_f;
    int m_numIntervals;
};

#endif /* defined(__FinancialSamples__SimpsonsIntegration__) */

//
//   SimpsonsIntegration.cpp

#include "SimpsonsIntegration.h"

#include "MathFunction.h"

#include <iostream>
#include <cmath>

using std::cout;
using std::endl;

namespace  {
    const int DEFAULT_NUM_INTERVALS = 100;
}

SimpsonsIntegration::SimpsonsIntegration(MathFunction<double> &f)
: m_f(f),
  m_numIntervals(DEFAULT_NUM_INTERVALS)
{
}

SimpsonsIntegration::SimpsonsIntegration(const SimpsonsIntegration &p)
: m_f(p.m_f),
  m_numIntervals(p.m_numIntervals)
{
}
```

```cpp
SimpsonsIntegration::~SimpsonsIntegration()
{
}

SimpsonsIntegration &SimpsonsIntegration::operator=(const
SimpsonsIntegration &p)
{
    if (this != &p)
    {
        m_f = p.m_f;
        m_numIntervals = p.m_numIntervals;
    }
    return *this;
}

double SimpsonsIntegration::getIntegral(double a, double b)
{
    double S = 0;
    double intSize = (b - a)/m_numIntervals;
    double x = a;

    for (int i=0; i<m_numIntervals; ++i)
    {
        S += (intSize / 6) * ( m_f(x) + m_f(x+intSize) + 4* m_f ((x +
        x+intSize)/2) );
        x += intSize;
    }
    return S;
}

void SimpsonsIntegration::setNumIntervals(int n)
{
    m_numIntervals = n;
}

// Example function

namespace  {
```

```
    class F2 : public MathFunction<double>
    {
    public:
        ~F2();
        double operator()(double x);
    };

    F2::~F2()
    {
    }

    double F2::operator()(double x)
    {
        return sin(x);
    }

}

int main()
{
    F2 f;
    SimpsonsIntegration si(f);
    double integral = si.getIntegral(0.5, 2.5);
        cout << " the integral of the function is " << integral << endl;

    si.setNumIntervals(200);
    integral = si.getIntegral(0.5, 2.5);
        cout << " the integral of the function with 200 intervals is " <<
        integral << endl;
    return 0;
}
```

Running the Code

The code displayed in Listing 10-3 was tested using the function $f(x) = \sin x$. The compiler used was gcc on Mac OS X and Windows. The program was tested on both platforms, generating identical results.

After compiling the class `SimpsonsIntegration`, you can run the application and observe output similar to the following:

```
./simpsonsIntegration
the integral of the function is 1.67873
the integral of the function with 200 intervals is 1.67873
```

As you can observe from these results, the accuracy of the solution with 100 intervals is similar to the accuracy for 200 intervals. This shows that 100 subdivisions are already enough to get very good results for this technique.

Conclusion

Integrating functions is one of the basic tasks in computational mathematics, due to the great importance of integration (also known as antiderivative) as a fundamental area of calculus. In the development of financial algorithms, there are also many situations where it is necessary to find quick solutions to problems that involve the evaluation of definite integrals.

In this chapter, you have learned a few C++ programming examples that explore some of the most common techniques for numerical integration. You have seen how integration methods such as the trapezoid and Simpson's rule can be applied to the task of finding the area under the curve for some preestablished functions.

The trapezoid method is the second important algorithm used to evaluate definite integrals. Given a general function, this method uses the function value at the extremes of the interval in order to define a trapezoid-based geometric approximation. You have seen some examples of how this strategy works, along with working code to implement the rule.

I have also discussed the well-known Simpson's method for definite integration. Here, the approximation to the curve is performed using a quadratic equation. You saw an example of how to use polynomial equations to achieve the desired accuracy. Using Simpson's method, you can perform integration with very good approximations, despite the fact that fewer subintervals may be necessary to read this accuracy.

Partial differential equations (PDEs) are another important mathematical tool for the financial software developer. It is important to understand how they work, as well as having tools to find solutions based on PDEs. In the next chapter, I discuss some important PDE-related techniques and their implementation in C++.

CHAPTER 11

Solving ODEs and PDEs

The solution of ODEs (ordinary differential equations) and PDEs (partial differential equations) is at the heart of many techniques used in the analysis of financial markets. Important analytical tools for derivative valuation such as the Black-Scholes model for stock options and other derivatives can be directly represented as differential equations. Such equations need to be regularly solved in order to determine the price of financial instruments traded in the global markets. This creates the need for high-performance code, capable of finding efficient solutions to these mathematical models.

Due to the large number of applications of ODEs and PDEs in science, engineering, and finance, several methods to solve them have been developed. In addition to the exact mathematical methods, capable of analyzing and finding the solution to differential equations, a software engineer also has to deal with purely computational approaches, as well as their implementation in C++.

Examples of differential equations of interest in finance include

- Thiele's differential equation: Used to determine fair prices for life insurance contracts

- Black-Scholes differential equation: Used to price options and related derivatives

- Market reserve differential equation

- Dynamic variations of portfolio optimization

- Merton's equations for utility optimization

- Along with several variations of these differential equations

Since the application of differential equations to financial problems is such a large area, in this chapter I am able to present only an overview of the methods most frequently used for their solution.

© Carlos Oliveira 2021

C. Oliveira, *Practical C++20 Financial Programming*, https://doi.org/10.1007/978-1-4842-6834-6_11

The programming examples discussed in this chapter cover a few particular aspects of ODE and PDE modeling and applications. Topics that you will explore include the following:

- Euler's method for ODEs: An algorithm that is simple to implement and can be applied directly to any first-order ODE.

- Runge-Kutta method: An improvement over the general ideas of Euler's algorithm, the Runge-Kutta method provides better stability and accuracy for the solution to ODEs.

- Black-Scholes equation: A general discussion of the Black-Scholes PDE and an overview of the forward method to solve this model.

Solving Ordinary Differential Equations

In this section, we will create a class to solve ODEs using Euler's method.

Solution

I start the discussion of differential equations with some methods for the numerical solution to ordinary differential equations. Before I can start with a first example, however, let's remember some of the relevant facts about ODEs.

An ordinary differential equation is an equation that includes the rate of change (derivative) with respect to a single variable in one or more of its terms. Given a differential equation, its *order* is defined as the maximum order of any of the derivatives included in the equation. The following are a few examples of ODEs:

$$x^3 \frac{d^2 y}{dx^2} + x \frac{dy}{dx} + y = x^5$$

$$x \frac{dy}{dx} + 4x^2 = y$$

Both equations involve the derivative of the variable y with respect to x. In the first equation, the derivative is applied twice, resulting in the term d^2y/dx^2, which means that it is a second-order ODE. The second equation contains only a first-order derivative with respect to x, making it a first-order ODE.

Standard equations (the ones that don't involve derivatives) usually have solutions that can be expressed as a single number. ODEs, however, include derivatives, and therefore their solutions are better described as being one or more functions, which together satisfy the conditions implied by the derivatives. For example, the following well-known differential equation describes Newton's law of gravity:

$$m\frac{d^2x}{dt^2} = -mg$$

The solution of such an equation is a general function describing the velocity and acceleration of an object. To find out a numeric solution to such a particular problem, you would need to supply one or more initial conditions that, when plugged into the general solutions, will provide an explicit value for x in the given equation.

As you have seen from the previous example, numerically solving an ODE involves working with initial conditions that can be substituted in the general function that solves the equation. As a consequence, numerical methods to solve ODEs (and PDEs) require the determination of initial conditions as a prerequisite to find their numerical solutions.

There are two main types of methods that can be used to solve differential equations. The first kind of solution is based on *symbolic methods*. Such methods use algebraic techniques, including the known rules of differentiation and integration, to simplify and derive a closed solution to a differential equation. Symbolic methods can be performed manually or by computers, and there is a class of software that was created specifically to perform such symbolic manipulations. Main examples include Mathematica, Maple, and Maxima, among others.

While symbolic methods are very useful in solving certain classes of ODEs and PDEs, many differential equations are too complex to be solved to a closed form using symbolic manipulation. Moreover, such symbolic techniques are very specialized and are normally used only during the modeling and exploration phases, when the engineer or mathematician is creating a model based on differential equations. For these reasons, symbolic techniques are mostly confined to specialized software packages, rather than being used as libraries for general-purpose languages.

The second class of techniques for solving differential equations is based on numerical algorithms. These algorithms are more general in the sense that they can be applied to any differential equation, as long as some basic requirements are met. Moreover, many common differential equations have no known closed solutions, and in such cases, numerical methods are the only ones available. Because numerical methods for ODEs and PDEs can be implemented using standard programming techniques, they are commonly used as part of mathematical libraries for programming languages such as FORTRAN and C++.

Euler's Method

The first numerical algorithm for ODEs you will learn about is a simple technique called Euler's method, which is based on the successive evaluation of the desired ODE at predetermined steps. Starting from a given initial condition, Euler's method tries to find the next value of the differential equation, using approximation formulas that are applied at predetermined intervals.

The idea of Euler's method is to correct possible errors in the evaluation of the ODE when starting from the given initial condition. For example, suppose that you want to evaluate an ODE at desired point c, when starting from initial condition x_0. To make the argument simpler, assume that $x_0 \leq c$, although the same ideas are valid in the other direction. To solve the ODE, the idea of this algorithm is to perform the evaluation in N steps, where N is a given parameter. As a consequence, the step size is given by

$$h = \frac{c - x_0}{N}.$$

Let's assume that the differential equation can be represented as a first-order ODE in the following general form:

$$y' = f(x,y)$$

Also, the initial condition (x_0, y_0) is known.

In general terms, at each step (given by the value h), Euler's algorithm will try to determine the correct value of the solution for the ODE for that small step size. The biggest problem, however, is that the solution to the differential equation is not known in an explicit way, so the algorithm has to guess a particular value for each step. Since the step size is a small interval, a possible way to guess the value of the function is to approximate it using a straight line. If we call y_t the value of the function at step t, then this leads to the following approximation for y_t:

$$y_t = y_{t-1} + h\frac{f\left(x_{t-1}, y_{t-1}\right) + f\left(x_{t-1} + h, y + hf\left(x_{t-1}, y_{t-1}\right)\right)}{2}$$

In other words, the algorithm takes the mean value of the linear approximation between the previous point and the next point, as the next approximation to the small value between y_{t-1} and y_t.

Euler's algorithm is a simple example of what is known as a predictor-corrector algorithm. Such methods work by predicting where a function might be in the subsequent iteration, which in this case is performed using a linear approximation. The next step is to correct this prediction, in this case by taking the average value. The same strategy is repeated in many other algorithms, although with more complex approximation schemes.

One of the biggest issues when using Euler's algorithm is controlling for errors in the result. As you have seen, the step size for Euler's method is one of the input parameters for its implementation and indicates the frequency with which we want to update the results of the differential equation. The finer grained the steps we take in this evaluation process, the closer to the real function we get. On the other hand, two problems occur when we increase the number of steps in the ODE evaluation. First, there is the increase in running time due to the additional calculations that become necessary. Second, and of even more concern, is the fact that by increasing the number of steps, you might be increasing the numeric errors that are inevitable when doing calculations on a computer. Solving these precision problems leads to the development of other methods, as you see in the next section.

Complete Code

Euler's method, as described in the previous section, is implemented in the EulersMethod class displayed in Listing 11-1. The important method in this class is solve(), which receives as parameters the number of steps, the initial values x_0 and y_0, and the target point c.

Listing 11-1. Implementation for Euler's Method for Solving ODEs

```
//
// EulersMethod.h

#ifndef __FinancialSamples__EulersMethod__
#define __FinancialSamples__EulersMethod__
```

```
template <class T>
class MathFunction;

class EulersMethod {
public:
    EulersMethod(MathFunction<double> &f);
    EulersMethod(const EulersMethod &p);
    ~EulersMethod();
    EulersMethod &operator=(const EulersMethod &p);

    double solve(int n, double x0, double y0, double c);
private:
    MathFunction<double> &m_f;
};

#endif /* defined(__FinancialSamples__EulersMethod__) */

//
//  EulersMethod.cpp

#include "EulersMethod.h"

#include "MathFunction.h"

#include <iostream>

using std::cout;
using std::endl;

EulersMethod::EulersMethod(MathFunction<double> &f)
: m_f(f)
{
}

EulersMethod::EulersMethod(const EulersMethod &p)
: m_f(p.m_f)
{
}
```

```cpp
EulersMethod::~EulersMethod()
{
}

EulersMethod &EulersMethod::operator=(const EulersMethod &p)
{
    if (this != &p)
    {
        m_f = p.m_f;
    }
    return *this;
}

double EulersMethod::solve(int n, double x0, double y0, double c)
{
    // problem :    y' = f(x,y) ;   y(x0) = y0

    auto x = x0;
    auto y = y0;
    auto h = (c - x0)/n;

    cout << " h is " << h << endl;

    for (int i=0; i<n; ++i)
    {
        double F = m_f(x, y);
        auto G = m_f(x + h, y + h*F);

        cout << " F: " << F << " G: " << G << endl;

        // update values of x, y
        x += h;
        y += h * (F + G)/2;
        cout << " x: " << x << " y: " << y << endl;

    }

    return y;
}
```

```
/// -----

class EulerMethSampleFunc : public MathFunction<double> {
public:
    double operator()(double x) { return x; } // not used
    double operator()(double x, double y);
};

double EulerMethSampleFunc::operator()(double x, double y)
{
    return  3 * x + 2 * y + 1;
}

int main()
{
    EulerMethSampleFunc f;
    EulersMethod m(f);
    double res = m.solve (100, 0, 0.25, 2);
    cout << " result is " << res << endl;
    return 0;
}
```

Running the Code

You can generate a binary executable from the source code in Listing 11-1 using any standards-compliant compiler such as gcc. Then, you can execute the code to get sample results such as the following for the sample equation $f(x) = 3x + 2y + 1$:

```
./eulersMethod
h is 0.02
 F: 1.5      G: 1.62     x: 0.02 y: 0.2812
 F: 1.6224  G: 1.7473   x: 0.04 y: 0.314897
 F: 1.74979 G: 1.87979 x: 0.06 y: 0.351193
 F: 1.88239 G: 2.01768 x: 0.08 y: 0.390193
 // ...
 F: 137.938 G: 143.515 x: 1.94 y: 68.4034
 F: 143.627 G: 149.432 x: 1.96 y: 71.334
```

```
 F: 149.548 G: 155.59  x: 1.98 y: 74.3854
 F: 155.711 G: 161.999 x: 2    y: 77.5625
result is 77.5625
```

Notice that the solution tests the approximation for two cases: when the number of intervals is 100 (the default) and when necessary.

Runge-Kutta Method for Solving ODEs

In this section, we will implement the Runge-Kutta method for solving ODEs.

Solution

In the last section, you saw how to use Euler's method to solve ODEs, a technique that iterates through a series of steps while computing an approximation to the desired differential equations. A problem with Euler's method, however, is its slow convergence. Due to the first-order approximation used, the method requires a large number of steps if any accuracy is desired. On the other hand, it is also difficult to avoid error propagation when the number of steps increases, which makes it difficult to improve the accuracy of this method.

To reduce some of the problems inherent in Euler's method, other strategies have been devised. The way these methods try to overcome such limitations is to use better approximations for each step of the algorithm. This way, it is possible to use fewer steps overall to find the desired solution. Also, the improved approximation makes it possible to reduce computational errors incurred during a single step.

One of the most popular of such improved algorithms for the solution of ODEs is called the Runge-Kutta method. Compared to Euler's method, the Runge-Kutta method uses a different approximation scheme for each new step of the algorithm, which guarantees higher accuracy. As a consequence, you will also have faster convergence when using the Runge-Kutta method.

As before, assume that we are given a first-order differential equation with relation to the x variable:

$$y' = f(x,y)$$

The initial condition (x_0, y_0) is known, and the goal is to calculate the value of the differential equation at some point c. If we define as N the number of steps, the step size can be given as

$$h = \frac{c - x_0}{N}.$$

The well-known Taylor method from calculus can be used to compute the approximation to a function given its derivatives. The approximation found using the second-order Taylor approximation will give a more accurate result than the linear approximation used in Euler's algorithm. The formulas used in the original Runge-Kutta algorithm are the following:

$$x_{t+1} = x_t + h$$

$$y_{t+1} = y_t + hf\left(x_t + \frac{h}{2}, y_t + \frac{h}{2}f(x_t, y_t)\right)$$

If you employ higher-order approximations derived using the Taylor method, you can get even more precise results. The most common of such approximations is the fourth-order Runge-Kutta method. In this case, the formula for y_{t+1} is given by

$$k_1 = hf(x_t, y_t)$$

$$k_2 = hf\left(x_t + \frac{h}{2}, y_t + \frac{k_1}{2}\right)$$

$$k_3 = hf\left(x_t + \frac{h}{2}, y_t + \frac{k_2}{2}\right)$$

$$k_4 = hf(x_t + h, y_t + k_3)$$

$$y_{t+1} = y_t + \frac{1}{6}(k_1 + 2k_2, 2k_3 + k_4)$$

This method offers good results in terms of fast approximation and is appropriate to solve most ODE problems. The implementation is relatively straightforward, as shown in the code that follows.

The updated algorithm can be seen in the function solve, which can be written as follows:

```
double RungeKuttaODEMethod::solve(int n, double x0, double y0, double c)
{
    auto x = x0;
    auto y = y0;
    auto h = (c - x0)/n;

    for (int i=0; i<n; ++i)
    {
        auto k1 = h * m_f(x, y);
        auto k2 = h * m_f(x + (h/2), y + (k1/2));
        auto k3 = h * m_f(x + (h/2), y + (k2/2));
        auto k4 = h * m_f(x + h, y + k3);

        x += h;
        y += ( k1 + 2*k2 + 2*k3 + k4)/6;
    }
    return y;
}
```

Complete Code

Listing 11-2 presents the Runge-Kutta method for solving ODEs. The code organization is similar to what I used for Euler's method in the previous section. The main difference resides in the way the next step is defined, which uses the equations based on the Taylor method as explained previously.

Listing 11-2. Implementation of the Runge-Kutta Method to Solve ODEs

```
//
// RungeKuttaODEMethod.h

#ifndef __FinancialSamples__RungeKuttaODEMethod__
#define __FinancialSamples__RungeKuttaODEMethod__
```

```cpp
template <class T>
class MathFunction;

class RungeKuttaODEMethod {
public:
    RungeKuttaODEMethod(MathFunction<double> &f);
    RungeKuttaODEMethod(const RungeKuttaODEMethod &p);
    ~RungeKuttaODEMethod();
    RungeKuttaODEMethod &operator=(const RungeKuttaODEMethod &p);
    double solve(int n, double x0, double y0, double c);
private:
    MathFunction<double> &m_f;
};

#endif /* defined(__FinancialSamples__RungeKuttaODEMethod__) */

//
//  RungeKuttaODEMethod.cpp

#include "RungeKuttaODEMethod.h"

#include "MathFunction.h"

#include <iostream>

using std::cout;
using std::endl;

RungeKuttaODEMethod::RungeKuttaODEMethod(MathFunction<double> &f)
: m_f(f)
{
}

RungeKuttaODEMethod::RungeKuttaODEMethod(const RungeKuttaODEMethod &p)
: m_f(p.m_f)
{
}

RungeKuttaODEMethod::~RungeKuttaODEMethod()
{
}
```

```
RungeKuttaODEMethod &RungeKuttaODEMethod::operator=(const
RungeKuttaODEMethod &p)
{
    if (this != &p)
    {
        m_f = p.m_f;
    }
    return *this;
}

double RungeKuttaODEMethod::solve(int n, double x0, double y0, double c)
{
    auto x = x0;
    auto y = y0;
    auto h = (c - x0)/n;

    for (int i=0; i<n; ++i)
    {
        auto k1 = h * m_f(x, y);
        auto k2 = h * m_f(x + (h/2), y + (k1/2));
        auto k3 = h * m_f(x + (h/2), y + (k2/2));
        auto k4 = h * m_f(x + h, y + k3);

        x += h;
        y += ( k1 + 2*k2 + 2*k3 + k4)/6;
    }
    return y;
}

/// -----

class RKMethSampleFunc : public MathFunction<double> {
public:
    double operator()(double x) { return x; } // not used
    double operator()(double x, double y);
};
```

```
double RKMethSampleFunc::operator()(double x, double y)
{
    return  3 * x + 2 * y + 1;
}

int main()
{
    RKMethSampleFunc f;
    RungeKuttaODEMethod m(f);
    double res = m.solve (100, 0, 0.25, 2);
    cout << " result is " << res << endl;
    return 0;
}
```

Running the Code

The code in Listing 11-2 was compiled and tested using gcc. It should work, however, using any standards-compliant compiler. The test code is in the main function, which runs the algorithm using the differential equation $y' = 3x + 2y + 1$. The results can be compared with what was achieved with Euler's method discussed previously.

```
./rungeKutta
 x: 0.02 y: 0.281216
 x: 0.04 y: 0.314931
 x: 0.06 y: 0.351245
 x: 0.08 y: 0.390266
 // ...
 x: 1.9  y: 62.9518
 x: 1.92 y: 65.6582
 x: 1.94 y: 68.4763
 x: 1.96 y: 71.4107
 x: 1.98 y: 74.466
 x: 2    y: 77.6472
result is 77.6472
```

The sample output shows the convergence of the algorithm, with 100 iterations. You can see the complete results displayed in Figure 11-1.

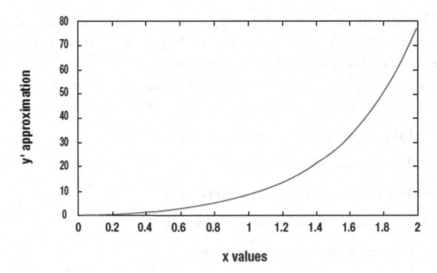

Figure 11-1. *Successive steps of the Runge-Kutta algorithm for the previous example, with N=100*

Solving the Black-Scholes Equation

Create a C++ class to solve the Black-Scholes equation using the forward method.

Solution

The Black-Scholes equation is one of the best-known methods to price derivatives. It was developed in the 1970s to provide a better model of European-style options, but since then, the basic model has been extended and tested on multiple derivatives markets. While the original assumptions of the Black-Scholes equation are not exactly respected in the real markets, the model works as an excellent analytical tool to price instruments that present volatile behavior as observed in the stock market.

Remember that an option is a contract that allows the holder to buy (or sell) units of a stock at a particular price in a given time in the future. For example, a call option on MSFT at $30 for July of the next year gives its owner the right (but not the duty) to buy MSFT for the price of $30 in July, irrespective of the real price at that date. Therefore, if MSFT stock price is significantly higher than $30, this operation will result in a profit. At $30 or lower prices, however, this option will result in a loss. Similarly, you can do the same analysis for a put, which is the right to sell a stock at the given price in the future.

A call produces higher profits when the price for the underlying asset rises, with a fixed maximum loss. On the other hand, a put produces higher profits when prices for the underlying asset decrease, also with a fixed maximum loss.

The Black-Scholes model defines what should be the present value of a call option (a similar analysis works for put options). It considers the following input values:

- S: The price of the underlying instrument

- K: The strike price for the option

- T: The remaining time for the option contract

- V: The current volatility of the underlying asset

- r: The current interest rate on deposits

Using this information, the Black-Scholes model concludes that the relationship between the current price of the option and the input variables is given by a PDE, as follows:

$$\frac{\partial C(S,t)}{\partial t} = rS\frac{\partial C(S,t)}{\partial S} + \frac{V^2 S^2}{2}\frac{\partial^2 C(S,t)}{\partial S^2} - rC(S,t)$$

Here, the implicit function $C(S,t)$ is the price or the derivative, which depends on the underlying price S and the time t.

There are several ways to solve differential equations like the previous one. The one you will use in this section is called the forward method. The general idea of solving PDEs is not very different from what you have seen for ODEs: take small steps toward the desired point that needs to be calculated, and evaluate the differential equation at these intermediate points using some kind of approximation. Unlike ODEs, which have only one dimension, however, PDEs have partial derivatives over two or more variables. In this case, we have partial derivatives in relation to the variables t and S. When this happens, the approximations become more complicated, because one needs to determine the shape of the small intervals at which the PDE will be evaluated. For example, the simplest scheme would be to divide the two-dimensional space into small rectangles and approximate over these small elements. Depending on the class of PDE, one can come up with more complicated and more precise ways to divide the domain and approximate the true value of the partial equation.

The forward difference method is an extension of Euler's method for PDEs. For the two-dimensional case, it can be used to divide the domain into rectangular elements.

For this method to work, you can assume that the stock price domain (S) varies between 0 and MaxS, a constant number that is in practice much higher than the desired strike price. The time domain varies from 0 (current date) to the future time (T) in which the option contract expires. Using these assumptions, the next task is to derive the equations that approximate the PDE at the next step, which can be done again using the Taylor method discussed earlier.

The initial conditions for the forward method are derived from the gain-loss equation at expiration. At that time, the value of an option is the positive difference between the stock price and the strike price, that is, max($S-K$,0). Therefore, the steps of the algorithm are inverted in the time dimension, starting from time T and moving backward to the present.

The resulting algorithm is presented in member function solve of class BlackScholesForwardMethod. The initial part of the function calculates terms of the equation that are unchanged over time. The three main factors are stored in the vectors a, b, and c.

$$a_n = \frac{1}{2}\left(nrdt - \left(nV\right)^2 dt\right)$$

$$b_n = 1 - rdt + \left(nV\right)^2 dt$$

$$c_n = \frac{1}{2}\left(nrdt + \left(nV\right)^2 dt\right)$$

The next step is to initialize the process using the given initial conditions. The calculated prices are stored in the two-dimensional vector u, which is initialized using the prices at expiration time. Then, the algorithm proceeds to compute the values for each of the time periods starting from expiration. At each day, starting from the day before expiration, the price of the option is calculated for each small increase in the underlying price. The option price for underlying value S depends on the price of the next day for values $S - dS$, S, and $S + dS$, where dS is a small increase in price, as determined by the parameter nx. Therefore, we have

$$u_{t,n} = a_n u_{t,n-1} + b_n u_{t,n} + c_n u_{t,n+1}$$

Complete Code

Listing 11-3 displays the complete implementation for the Black-Scholes forward method. You will find the code in class BlackScholesForwardMethod, along with a sample of its use in function main().

Listing 11-3. Black-Scholes Forward Method Implementation

```
//
// BlackScholesForwardMethod.h

#ifndef __FinancialSamples__BlackScholesForwardMethod__
#define __FinancialSamples__BlackScholesForwardMethod__

#include <vector>

class BlackScholesForwardMethod {
public:
    BlackScholesForwardMethod(double expiration, double maxPrice, double
    strike, double intRate);
    BlackScholesForwardMethod(const BlackScholesForwardMethod &p);
    ~BlackScholesForwardMethod();
    BlackScholesForwardMethod &operator=(const BlackScholesForwardMethod &p);

    std::vector<double> solve(double volatility, int nx, int timeSteps);
private:
    double m_expiration;
    double m_maxPrice;
    double m_strike;
    double m_intRate;
};

#endif /* defined(__FinancialSamples__BlackScholesForwardMethod__) */

//
// BlackScholesForwardMethod.cpp

#include "BlackScholesForwardMethod.h"
```

```cpp
#include <cmath>
#include <algorithm>
#include <vector>
#include <iostream>
#include <iomanip>

using std::vector;
using std::cout;
using std::endl;
using std::setw;

BlackScholesForwardMethod::BlackScholesForwardMethod(double expiration,
double maxPrice,

                                                     double strike, double
                                                     intRate)
: m_expiration(expiration),
  m_maxPrice(maxPrice),
  m_strike(strike),
  m_intRate(intRate)
{
}

BlackScholesForwardMethod::BlackScholesForwardMethod(const
BlackScholesForwardMethod &p)
: m_expiration(p.m_expiration),
  m_maxPrice(p.m_maxPrice),
  m_strike(p.m_strike),
  m_intRate(p.m_intRate)
{
}

BlackScholesForwardMethod::~BlackScholesForwardMethod()
{
}

BlackScholesForwardMethod &BlackScholesForwardMethod::operator=(const
BlackScholesForwardMethod &p)
{
```

```cpp
    if (this != &p)
    {
        m_expiration = p.m_expiration;
        m_maxPrice = p.m_maxPrice;
        m_strike = p.m_strike;
        m_intRate = p.m_intRate;
    }
    return *this;
}

vector<double> BlackScholesForwardMethod::solve(double volatility, int nx,
int timeSteps)
{
    double dt = m_expiration /(double)timeSteps;
    double dx = m_maxPrice /(double)nx;

    vector<double> a(nx-1);
    vector<double> b(nx-1);
    vector<double> c(nx-1);

    int i;
    for (i = 0; i < nx - 1; i++)
    {
        b[i] = 1.0 - m_intRate * dt - dt * pow(volatility * (i+1), 2);
    }

    for (i = 0; i < nx - 2; i++)
    {
        c[i] = 0.5 * dt * pow(volatility * (i+1), 2) + 0.5 * dt * m_intRate
        * (i+1);
    }

    for (i = 1; i < nx - 1; i++)
    {
        a[i] = 0.5 * dt * pow(volatility * (i+1), 2) - 0.5 * dt * m_intRate
        * (i+1);
    }

    vector<double> u((nx-1)*(timeSteps+1));
```

```
    double u0 = 0.0;
    for (i = 0; i < nx - 1; i++)
    {
        u0 += dx;
        u[i+0*(nx-1)] = std::max(u0 - m_strike, 0.0);
    }

    for (int j = 0; j < timeSteps; j++)
    {
        double t = (double)(j) * m_expiration /(double)timeSteps;

        double p = 0.5 * dt * (nx - 1) * (volatility*volatility * (nx-1) +
        m_intRate)
        * (m_maxPrice-m_strike * exp(-m_intRate*t ) );

        for (i = 0; i < nx - 1; i++)
        {
            u[i+(j+1)*(nx-1)] = b[i] * u[i+j*(nx-1)];
        }
        for (i = 0; i < nx - 2; i++)
        {
            u[i+(j+1)*(nx-1)] += c[i] * u[i+1+j*(nx-1)];
        }
        for (i = 1; i < nx - 1; i++)
        {
            u[i+(j+1)*(nx-1)] += a[i] * u[i-1+j*(nx-1)];
        }
        u[nx-2+(j+1)*(nx-1)] += p;
    }

    return u;
}

int main()
{
    auto strike = 5.0;
    auto intRate = 0.03;
    auto sigma = 0.50;
```

```
auto t1 = 1.0;
auto numSteps = 11;
auto numDays = 29;
auto maxPrice = 10.0;

BlackScholesForwardMethod bsfm(t1, maxPrice, strike, intRate);
vector<double> u = bsfm.solve(sigma, numSteps, numDays);

double minPrice = .0;
for (int  i=0; i < numSteps-1; i++)
{
    double s = ((numSteps-i-2) * minPrice+(i+1)*maxPrice)/ (double)
    (numSteps-1);
    cout << "  " << s << "  " << u[i+numDays*(numSteps-1)] << endl;
}
return 0;
}
```

Running the Code

To test the code displayed in Listing 11-3, you can build it using any standards-compliant compiler. I run this code using the llvm C++ compiler, with the following results:

./blackScholes
```
 1   0.000452875
 2   0.0148578
 3   0.109172
 4   0.361706
 5   0.784941
 6   1.34918
 7   2.016
 8   2.75175
 9   3.53055
10   4.33362
```

This result means, for example, that 29 days from expiration, a call option with strike price $5 and volatility 0.5 would be valued at $1.3 when the price of the underlying is $6. Notice that you can use this code to calculate prices for each price level ranging from $1 up to $10. You can also modify the code to compute option prices for more expensive stocks.

Conclusion

Solving differential equations is a big part of financial analysis techniques. Such techniques are used in many areas where the price of assets is determined by complex differential equations such as the Black-Scholes model, which is the main technique used by banks to price equity derivatives and related investments.

In this chapter, I introduced you to the topic of numerical solutions of differential equations. Although this is a large area that cannot be easily covered in a single chapter, it is useful to understand the basic techniques and how they are employed in the field of financial programming.

Euler's method for ODEs is the first method discussed. Its main idea is to perform several steps, where each step approximates the result of the differential equation. The second method, the Runge-Kutta algorithm, is an improvement on this general strategy, using higher-order Taylor approximations that make the algorithm more accurate and avoid some of the weaknesses of Euler's method. You have seen how to implement both algorithms in C++, with test data that demonstrates their convergence.

The Black-Scholes equation is one of the most important mathematical models in modern finance. While there are several robust and efficient algorithms for its solution, I present a simple method based on forward differences. You have seen how the general solution strategy works and how it can be efficiently implemented in C++.

Finding solutions to equations that model market behavior as viewed in this chapter is generally the beginning of a process of data analysis. Another step is to find the best solution that meets a particular investment goal. For this purpose, a number of optimization techniques have been developed. In the next chapter, I present some general optimization methods that have been successfully used in the analysis of financial investments, along with their implementation in C++.

CHAPTER 12

Optimization

Optimization is a wide area that covers a large set of techniques used to find the minimum or maximum of a function over a predefined group of conditions. Optimization strategies are frequently employed in several areas of financial engineering such as portfolio optimization and as such should be part of the basic skill set of financial developers.

In this chapter, we discuss programming examples that explore a few of the implement aspects of optimization algorithms. We start with a concise explanation of some techniques and how they are typically implemented in C++. Topics covered in this chapter include the following:

- Optimization concepts: Basic concepts on optimization and how it is used as a common step of algorithms for financial applications.

- Linear programming models: The basics of linear optimization models, with common assumptions and how the results can be interpreted. You will also learn how to create linear programming models for common problems.

- Solving linear models: You will learn about techniques and algorithms commonly used to solve linear programming models. In particular, you will learn how to employ a popular open source library to solve linear programming problems.

- Solving mixed-integer programming models: A common extension of linear programming is to require that one or more decision variables assume only integer values. This type of problem, called an integer programming problem, is frequently used whenever mutually exclusive choices are part of a linear model. You will also learn how to extend the linear programming class to solve such mixed-integer programming models.

© Carlos Oliveira 2021
C. Oliveira, *Practical C++20 Financial Programming*, https://doi.org/10.1007/978-1-4842-6834-6_12

Interfacing with a Linear Programming Solver

In this section, we create a generic class to solve a linear programming problem, given the objective function and constraints in matrix form.

Solution

Optimization is a mathematical technique used to find the maximum or minimum of a given function over a set of constraints. The methods currently used in optimization have started as a set of simple results from calculus, where a single function is subject to minimization or maximization. Nowadays, these techniques include complex models involving multiple linear and nonlinear components.

In financial engineering and economics, optimization is a tool used for purposes such as designing an optimal asset portfolio allocation or more widely to determine the best investment decision from a large set of asset classes. Due to its origins in the analysis of scarce resources and their optimum use, linear programming has been a favorite tool for economists and financial analysts—which shows why optimization is such a common technique in financial programming. Effectively, every time we need to make a decision on asset allocation given a large number of scenarios, optimization becomes a useful tool to help select the best decision.

In the code example in the section "Using LP Solver Libraries," we consider how to interface with existing libraries that can be used to solve a large class of optimization problems. To keep the discussion well contained, I employ an open source library called GLPK (Gnu Linear Programming Kit), which will be also used as a basis for future examples. GLPK is simple to use, but it is a C-based library only. This means that it provides no direct support for C++ high-level concepts such as classes, templates, and containers. Therefore, as part of the discussion, I will show you how to create a class that provides a basic C++ interface to GLPK and other solvers.

First, however, I give you some preliminary information about the kinds of problems that can be solved with an optimization engine, starting with linear programming. Then, I present some code that can be used to translate simple linear models into calls to the solver application programming interface (API).

Linear Programming Concepts

The first case of optimization that you will learn about is characterized by an *objective* represented as a linear function. This objective is then optimized over a set of linear functions, also called *constraints*. Such optimization problems are called linear programming (LP) problems, and they constitute an important class of mathematical models that have been widely used in disciplines such as financial analysis and economics.

Using a more formal (mathematical) definition, LP is the area of optimization that deals with the determination of the minimum or maximum value of a linear function over a set of linear constraints. Each constraint is of the form

$$\sum_{j=1}^{n} a_j x_j = a_1 + \ldots + a_n \leq b_i.$$

Similarly, the function that you want to optimize over (also known as *objective function*) is a linear function. This results in a problem that can be denoted in the following way:

$$minimize \sum_{j=1}^{n} c_j x_j$$

$$subject\ to \sum_{j=1}^{n} a_{ij} x_j \leq b_i \qquad for\ i \in \{1 \ldots m\}$$

$$x_j \geq 0 \quad for\ j \in \{1 \ldots n\}$$

In these equations, x_j is a variable, and a_{ij}, b_i, and c_j are constant values. These parameters are frequently provided as matrix A and two vectors, b and c. Due to its generic characteristics, this type of problem can assume several forms, depending on the exact value for the given coefficients, as well as if they are zero or nonzero. Also, variations of the problem involve the change of the $\leq$ relation to $\geq$ or $=$ in one or more equations. Finally, the problem can require the maximization, rather than the minimization, of the objective function. All these variants can be easily shown to be equivalent to each other, in the sense that it is possible to convert them to a particular form and use the same algorithm in their solution.

Solving an LP problem can be done with the help of a method called *simplex algorithm*. The basic approach of the simplex algorithm is to consider the geometric region defined by the constraints in a multidimensional space and start to visit the corners of this object in a well-defined way—until an optimal solution is found.

In essence, the mechanics of solving an LP problem are not very different from solving a sequence of linear systems, and a few strategies have been devised using this general strategy. The simplex algorithm, which is still the most common technique to solve LP problems, proceeds by defining a sequence of modified linear systems that are shown to be equivalent to the original while at the same time improving the value of the objective function. One of the advantages of the simplex algorithm is that its properties are well known—mathematical analyses of the simplex algorithm throughout the years have considered several important questions such as its convergence and performance.

While the operation of the simplex algorithm is not difficult to describe, the implementation of such an algorithm contains a lot of intricate edge cases. To avoid such issues, most frequently, you will be using an LP solver library, which has been especially designed to hide the complexities of the implementation. Essentially, a solver provides just a simple API so that users can call the algorithms, provide necessary data, and retrieve the results.

Using LP Solver Libraries

There are several commercial and free libraries that implement the simplex algorithm (and even a few more efficient algorithms for this problem). In this section, to give a flavor of how the process of modeling and LP works, we use a simple but well-maintained open source library called GLPK. With GLPK, it is possible to solve from medium- to relatively large-size LP problems (as well as a few other model variants such as mixed-integer programs).

To start using GLPK from C++, the first step is to download and compile the source code. You will find a version of this software in the Gnu website (at the time I checked, the URL was `www.gnu.org/software/glpk`). Unlike many math open source libraries, GLPK is very easy to compile and install. You need to decompress the file and build the library using the `configure` and `make` commands (these instructions work on UNIX systems, but you can download software such as Cygwin that will allow you to perform the same commands in Windows).

Once GLPK is installed, you can link to its library, `libglpk.a`, and make use of the functions that are exported by its API. On Windows systems, you can use the precompiled binary dll and lib files available on the GLPK website. You can also use the MingWin compiler for gcc on Windows. I present a class called `LPSolver` that is able to interface with the GLPK API. The following is the public part of the class declaration:

```
class LPSolver {
public:
    LPSolver(Matrix &m, const std::vector<double> &b, const
    std::vector<double> &c);
    LPSolver(const LPSolver &p);
    ~LPSolver();
    LPSolver &operator=(const LPSolver &p);

    enum ResultType {
        INFEASIBLE,
        FEASIBLE,
        ERROR
    };

    void setName(const std::string &s);
    bool isValid();
    void setMaximization();
    void setMinimization();
    ResultType solve(std::vector<double> &result, double &objValue);
// ...
};
```

First, an object of `LPSolver` type can be created if you pass a matrix A, a vector b, and a vector c to the constructor. These parameters are interpreted as the coefficients of the objective function as well as the constraints of the LP.

You can also give a descriptive name to the problem using the `setName` member function. Its implementation shows how a simple function in the GLPK looks.

```
void LPSolver::setName(const std::string &s)
{
    glp_set_prob_name(m_lp, s.c_str());
}
```

The API function is called `glp_set_prob_name`. The first parameter, as for most other functions in GLPK, is a pointer to the LP data structure. The second parameter, a string, is unique for this API call.

The `isValid` member function checks if the object has been properly initialized. The `setMaximization` and `setMinimization` member functions can be used to define the direction of the optimization.

Finally, the `solve` member function performs the optimization algorithm. This is done with a call to GLPK, where the `glp_simplex` function is used to do the hard work. After the optimization is finished, the algorithm collects the result of the objective function and the value of each variable for this optimal solution.

```
LPSolver::ResultType LPSolver::solve(std::vector<double> &result, double
&objValue)
{
    glp_simplex(m_lp, NULL);

    result.resize(m_M, 0);
    objValue = glp_get_obj_val(m_lp);

    for (int i=0; i<m_M; ++i)
    {
        result[i] = glp_get_col_prim(m_lp, i+1);
    }
    return LPSolver::FEASIBLE;
}
```

Finally, an example LP is used to test the `LPSolver` class. In this example, I provided objective function coefficients equal to 10, 6, and 4. The right-hand side of the constraints is also provided as a vector. Finally, the constraints of the problem are given in the form of `Matrix` object A.

Complete Code

Listing 12-1 displays the complete listing for the LP solver described in the previous section. An example of the class `LPSolver` is given in the `main` function at the end of the listing.

Listing 12-1. Class LPSolver Header and Implementation

```
//
// LPSolver.h

#ifndef __FinancialSamples__Glpk__
#define __FinancialSamples__Glpk__

#include <vector>
#include <string>

#include "Matrix.h"

struct glp_prob;

class LPSolver {
public:
    LPSolver(Matrix &A, const std::vector<double> &b,
             const std::vector<double> &c);
    LPSolver(Matrix &A, const std::vector<double> &b,
             const std::vector<double> &c,
             const std::string &probname);
    LPSolver(const LPSolver &p);
    ~LPSolver();
    LPSolver &operator=(const LPSolver &p);

    enum ResultType {
        INFEASIBLE,
        FEASIBLE,
        ERROR
    };

    virtual ResultType solve(std::vector<double> &result, double
    &objValue);
    void setName(const std::string &s);
    bool isValid();
    void setMaximization();
    void setMinimization();
```

```cpp
private:
    size_t m_M;
    size_t m_N;
    std::vector<double> m_c;
    std::vector<double> m_b;
    Matrix m_A;
    glp_prob *m_lp;

    void initProblem(size_t M, size_t N);
    void setRowBounds();
    void setColumnCoefs();
protected:
    glp_prob *getLP();
    int getNumCols();
    int getNumRows();
};

#endif /* defined(__FinancialSamples__Glpk__) */

//
//  LPSolver.cpp

#include "LPSolver.h"

#include <glpk.h>

#include <iostream>

using std::vector;
using std::string;
using std::cout;
using std::endl;

LPSolver::LPSolver(Matrix &m, const vector<double> &b, const vector<double> &c)
: m_M(m.numRows()),
  m_N(m[0].size()),
  m_c(c),
  m_b(b),
  m_A(m),
```

```
    m_lp(glp_create_prob())
{
    initProblem(m_M, m_N);
}

LPSolver::LPSolver(Matrix &m, const std::vector<double> &b,
              const std::vector<double> &c,
              const std::string &probname)
: m_M(m.numRows()),
  m_N(m[0].size()),
  m_c(c),
  m_b(b),
  m_A(m),
  m_lp(glp_create_prob())
{
    initProblem(m_M, m_N);
    glp_set_prob_name(m_lp, probname.c_str());
}

LPSolver::LPSolver(const LPSolver &p)
: m_M(p.m_M),
  m_N(p.m_N),
  m_c(p.m_c),
  m_b(p.m_b),
  m_A(p.m_A),
  m_lp(glp_create_prob())
{
    initProblem(m_M, m_N);
}

// performs necessary initialization of the given values
void LPSolver::initProblem(size_t M, size_t N)
{
    if (!m_lp) return;

    setRowBounds();
    setColumnCoefs();
```

```
    vector<int> I, J;
    vector<double> V;

    // indices in GLPK start on 1
    I.push_back(0);
    J.push_back(0);
    V.push_back(0);
    for (int i=0; i<M; ++i)
    {
        for (int j=0; j<N; ++j)
        {
            I.push_back(i+1);
            J.push_back(j+1);
            V.push_back(m_A[i][j]);
        }
    }
    glp_load_matrix(m_lp, (int)(m_M * m_N), &I[0], &J[0], &V[0]);
}

LPSolver::~LPSolver()
{
    glp_delete_prob(m_lp);
}

LPSolver &LPSolver::operator=(const LPSolver &p)
{
    if (this != &p)
    {
        m_M = p.m_M;
        m_N = p.m_N;
        m_c = p.m_c;
        m_b = p.m_b;
        m_A = p.m_A;
        m_lp = glp_create_prob();
        initProblem(m_M, m_N);
    }
```

```cpp
    return *this;
}

void LPSolver::setName(const std::string &s)
{
    glp_set_prob_name(m_lp, s.c_str());
}

bool LPSolver::isValid()
{
    return m_lp != NULL;
}

void LPSolver::setMaximization()
{
    glp_set_obj_dir(m_lp, GLP_MAX);
}

void LPSolver::setMinimization()
{
    glp_set_obj_dir(m_lp, GLP_MIN);
}

void LPSolver::setRowBounds()
{
    glp_add_rows(m_lp, (int)m_M);
    for (int i=0; i<m_M; ++i)
    {
        glp_set_row_bnds(m_lp, i+1, GLP_UP, 0.0, m_b[i]);
    }
}

void LPSolver::setColumnCoefs()
{
    glp_add_cols(m_lp, (int)m_N);
    for (int j=0; j<m_N; ++j)
    {
        glp_set_col_bnds(m_lp, j+1, GLP_LO, 0.0, 0.0);
```

```
            glp_set_obj_coef(m_lp, j+1, m_c[j]);
    }
}

LPSolver::ResultType LPSolver::solve(std::vector<double> &result, double
&objValue)
{
    glp_simplex(m_lp, NULL);
    result.resize(m_N, 0);
    objValue = glp_get_obj_val(m_lp);

    for (int j=0; j<m_N; ++j)
    {
        result[j] = glp_get_col_prim(m_lp, j+1);
    }
    return LPSolver::FEASIBLE;
}

glp_prob *LPSolver::getLP()
{
    return m_lp;
}

int LPSolver::getNumCols()
{
    return (int)m_N;
}

int LPSolver::getNumRows()
{
    return (int)m_M;
}

int main_lps()
{
    Matrix A(3);
    A[0][0] = 1;  A[0][1] = 1; A[0][2] = 1;
    A[1][0] = 10; A[1][1] = 2; A[1][2] = 4;
    A[2][0] = 2;  A[2][1] = 5; A[2][2] = 6;
```

```
vector<double> c = { 10, 6, 4 };
vector<double> b = { 100, 600, 300 };

LPSolver solver(A, b, c);
solver.setMaximization();
vector<double> results;
double objVal;
solver.solve(results, objVal);
for (int i=0; i<results.size(); ++i)
{
    cout << " x" << i << ": " << results[i];
}
cout << " max: " << objVal << endl;
return 0;
}
```

Running the Code

The code presented in Listing 12-1 can be compiled with a standards-compliant compiler, such as gcc or Visual Studio. Remember to add the GLPK library to the link step (in gcc, this is done with the -L and –l switches). The result of the program execution should be similar to the following:

./lpSolver
```
GLPK Simplex Optimizer, v4.54
3 rows, 3 columns, 9 non-zeros
*     0: obj =   0.000000000e+00   infeas =   0.000e+00 (0)
*     2: obj =   7.565217391e+02   infeas =   0.000e+00 (0)
OPTIMAL LP SOLUTION FOUND
 x0: 52.1739 x1: 39.1304 x2: 0 max: 756.522
```

Here, you see the first output of GLPK. By default, GLPK displays the best solutions and the number of iterations it has taken to achieve the results. You can see that after two iterations of the simplex algorithm, GLPK found a solution with an objective value equal to 756.

Solving Two-Dimensional Investment Problems

In this section, we use LP techniques to model and solve a financial product allocation decision problem for two investments with known returns.

Solution

One of the main uses of optimization techniques is in the support of investment decisions. In this respect, there are several concepts that can be optimized based on the known properties of investment classes. For a few types of investments, such as bonds, it is easier to determine the returns of the investment, as well as some basic information about the risk for that class of investments. This knowledge translates into more accurate models, particularly the ones that can be employed to optimize profits for the investor.

This section presents a very simple version of a decision support system modeled using linear programming. This problem shows the basic geometric process that is used to solve LPs (even the most complicated ones).

Consider the process of introducing two new financial products to the market in a large bank. The process is usually defined by a number of practical constraints on resources necessary for these investments. Suppose that the bank wants to add two classes of new products to the market: new bond-based products and new mortgage-backed derivatives products. The question is how many hours should be spent on the development of these new products. Let's call these two variables x and y. Since the bank unit receives payment from its clients based on the number of hours spent on these tasks, the goal is to maximize the amount paid per hours. For bonds, the cost is $5.3K per hour spent on that activity, while the price is $7.1K for derivatives.

In terms of constraints, the bank unit has to consider research expenditures costs. The cost of working with bonds is negative because its costs can be offset by other activities in that area. For derivatives, the full cost of research is considered. The maximum cost spent on marketing for these financial products is prorated by working hours too. Thus, it depends on the values of x and y. It is known that there is a constant of $3K for working hours on bond-related products, while the multiplier for derivatives work is $1K.

Finally, there is a limit on the amount of human resources available for these two tasks. While there are only six units of human resources allocated to these tasks, each hour of bond-related work is eight times more demanding than for derivatives. Notice also that the two variables in this example are clearly non-negative.

The result of these assumptions can be readily translated into the following LP model, which tries to maximize the expected return (profit) for the considered unit of the investment bank.

max $5.3x + 7.1y$ (maximize department results)

$-2.1x + y \leq 3.4$ (maximum research expenditure)

$3.1x + y \leq 4.3$ (maximum marketing expenditure)

$7.9x + y \leq 6$ (maximum number of employees needed)

$x, y \geq 0$ (working hours are always positive)

The model described previously has only two unknowns, x and y, and therefore can be readily plotted as seen in Figure 12-1. Being an inequality, each constraint results in a half-space that is defined by the equality line. For example, $3x + y \leq 4$ is the half-space defined by all points under the line $3x + y = 4$.

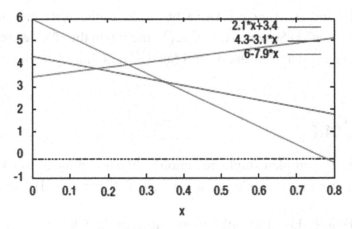

Figure 12-1. *Feasible set for the LP defined by inequalities shown earlier*

To find a solution for a two-dimensional LP model like the one described previously, you can concentrate on the intersections of all half-spaces defined by the constraints. The intersection is, by definition of the problem, contained in the first quadrant of the plot because it is known that $x \geq 0$ and $y \geq 0$. It is possible then to recognize the area contained in the intersection of all other half-spaces. The result is a polygonal area, whose border is defined by a set of lines derived from the given constraints.

To find a solution for such an LP, you just need to calculate the value of the objective function at each of the corners of the area defined by the constraints. The corner that gives the best value of the objective function is, by definition of the linear objective function, the best that can be found for the problem.

While the process described previously is easy to perform for two-dimensional problems, it becomes quite difficult to accomplish for higher dimensions. As the number of dimensions and constraints increases, the number of corners grows exponentially. It takes a more sophisticated algorithm (such as the simplex algorithm) to find the best corner of the multidimensional space that defines the optimum solution for the given LP.

To demonstrate how the problem is solved in practice, I show you the C++ implementation of the proposed two-dimensional LP. The class, called TwoDimensionalLPSolver, is a blueprint of how such a problem can be implemented using the LPSolver described in the previous section.

First, you need to create the model, which is described using a matrix A, a vector b (the right-hand side of the constraints), and a vector c (the cost vector). The necessary data is provided in the main function. Once the data is available, it can be used to create an object of class LPSolver. The solve() function in the LPSolver class will then perform any necessary data conversion and call the GLPK library to find the optimum solution.

Complete Code

Listing 12-2 gives the complete implementation for the two-dimensional LP solver. Function main() presents a sample use of the class TwoDimensionalLPSolver.

Listing 12-2. Header File and Implementation for the Class TwoDimensionalLPSolver

```
//
// TwoDimensionalLPSolver.h

#ifndef __FinancialSamples__TwoDimensionalLPSolver__
#define __FinancialSamples__TwoDimensionalLPSolver__

#include <vector>
```

```cpp
class TwoDimensionalLPSolver {
public:
    using Vector = std::vector<double>;

    TwoDimensionalLPSolver(const Vector &c, const Vector &A1, const Vector
    &A2, const Vector &b);
    TwoDimensionalLPSolver(const TwoDimensionalLPSolver &p);
    ~TwoDimensionalLPSolver();
    TwoDimensionalLPSolver &operator=(const TwoDimensionalLPSolver &p);

    bool solveProblem(Vector &results, double &objVal);
private:
    std::vector<double> m_c;
    std::vector<double> m_A1;
    std::vector<double> m_A2;
    std::vector<double> m_b;
};

#endif /* defined(__FinancialSamples__TwoDimensionalLPSolver__) */

//
//  TwoDimensionalLPSolver.cpp

#include "TwoDimensionalLPSolver.h"

#include "Matrix.h"
#include "LPSolver.h"

#include <iostream>

using std::vector;
using std::cout;
using std::endl;

TwoDimensionalLPSolver::TwoDimensionalLPSolver(const Vector &c, const
Vector &A1,

                                              const Vector &A2, const
                                              Vector &b)
: m_c(c),
  m_A1(A1),
```

```
    m_A2(A2),
    m_b(b)
{
}

TwoDimensionalLPSolver::TwoDimensionalLPSolver(const TwoDimensionalLPSolver &p)
: m_c(p.m_c),
  m_A1(p.m_A1),
  m_A2(p.m_A2),
  m_b(p.m_b)
{
}

TwoDimensionalLPSolver::~TwoDimensionalLPSolver()
{
}

TwoDimensionalLPSolver &TwoDimensionalLPSolver::operator=(const
TwoDimensionalLPSolver &p)
{
    if (this != &p)
    {
        m_c = p.m_c;
        m_A1 = p.m_A1;
        m_A2 = p.m_A2;
        m_b = p.m_b;
    }
    return *this;
}

bool TwoDimensionalLPSolver::solveProblem(Vector &res, double &objVal)
{
    int size = m_b.size();
    Matrix A(size, 2);
    for (int j=0; j<size; ++j)
    {
        A[j][0] = m_A1[j];
```

```
        A[j][1] = m_A2[j];
    }
    LPSolver solver(A, m_b, m_c);
    solver.setMaximization();
    return solver.solve(res, objVal) == LPSolver::ResultType::FEASIBLE;
}

int main()
{
    vector<double> A1 = { -2.1, 3.1, 7.9};
    vector<double> A2 = { 1, 1, 1 };
    vector<double> c = { 5.3, 7.1 };
    vector<double> b = { 3.4, 4.3, 6 };
    TwoDimensionalLPSolver solver(c, A1, A2, b);

    vector<double> results;
    double objVal;
    solver.solveProblem(results, objVal);
    cout << "objVal : " << objVal << endl;
    for (int i=0; i<results.size(); ++i)
    {
        cout << " x" << i << ": " << results[i];
    }
    cout << endl;
    return 0;
}
```

Running the Code

You can compile and run the provided code with your preferred standards-compliant compiler. I tested the code using gcc and GLPK optimizer version 4.54. The results are as follows:

./twoDimSolver
```
GLPK Simplex Optimizer, v4.54
3 rows, 2 columns, 6 non-zeros
```

```
*      0: obj = 0.000000000e+00 infeas = 0.000e+00 (0)
*      2: obj = 2.763788462e+01 infeas = 0.000e+00 (0)
OPTIMAL LP SOLUTION FOUND
objVal : 27.6379
 x0: 0.173077 x1: 3.76346
Program ended with exit code: 0
```

From the output listed in the example, you can see that the optimal solution was achieved at the vertex (0.173, 3.763), which corresponds to the intersection of equations $-2.1x + y = 3.4$ and $3.1x + y \leq 4$. At that point, the objective function has a value of 27.63, which can be interpreted as the profit achieved by the department in spending the given number of hours in the two financial products that were discussed earlier.

Creating Mixed-Integer Programming Models

Extend the LPSolver class so that it can deal with mixed-integer programming (MIP) problems, that is, LP problems where one or more variables are restricted to be integers.

Solution

After continuous LP problems, MIP problems are probably the most common type of optimization problem that practitioners need to deal with. In terms of modeling, the biggest difference between LP and MIP is that such problems have one or more decision variables that are required to be integer numbers—unlike LP problems, where all decision variables are continuous (normally real numbers).

Integer variables are ideal for cases where you need to make decisions that are exclusive within a small- to medium-size set. Moreover, these decision variables may be applicable to resources that are not divisible. For example, you can use such variables to decide on the number of local branches for a commercial bank or on the number of different stocks included in a portfolio. These are common examples of resources that can only be used in integer quantities.

A special type of integer variable is a binary decision variable, also called a 0–1 decision variable. These are variables that can assume only a 0 or 1 (all-or-nothing) value. They are the purest form of integer variable, because they allow one to decide between only two alternative choices. As you can expect, many MIP problems make use of binary variables as their primary way to reach an optimal decision.

In terms of techniques for problem solving, MIP problems are a lot more complicated than LP problems. While there are very efficient algorithms available to solve LP formulations, not all MIP problems are readily solvable by current computer algorithms. As a short explanation, this occurs because when a decision variable is an integer, it creates "jumps" in the objective function that make it much harder to search for the optimum solution. So, unlike LP problems where the optimal vertex of the set of feasible solutions can be quickly determined, MIP solvers need to spend much more time generating possible solutions and testing if they are optimal. This exponential explosion of options is the main reason why MIP problems are much more difficult to solve than LP problems.

Most LP libraries have been extended to deal with at least some forms of MIP. GLPK implements a generic algorithm for MIP solving, called branch-and-cut. With this algorithm, it is possible to solve small- to moderate-size MIPs to optimality. More complicated MIP problems, however, may not be solvable using this technique, depending on the structure of the required problem.

In this coding example, you will see how to extend the LPSolver class to deal with MIP problems, in addition to classical LP problems. In the next section, you will see an example of how to use the LPSolver class to model and solve a MIP problem.

The main reason I decided to inherit from LPSolver, instead of creating a new unrelated class, is that in terms of modeling, MIP problems are very close to LP problems. The only additional thing you need to do in the latter case is to tell which variables are integer or binary and call the right version of the function that solves and retrieves values found by GLPK.

In the MIPSolver class, this is implemented in the following way. First, there are two new member functions called setColBinary and setColInteger. These member functions can be used to tell GLPK that the variable in a given column is either integer or binary, respectively. Their implementations are straightforward and simply call the related C function in GLPK. For example:

```
void MIPSolver::setColBinary(int colNum)
{
    glp_set_col_kind(getLP(), colNum+1, GLP_BV);
}
```

The other part of the puzzle is to implement a new version of the solve member function. The new version supersedes the original version in LPSolver and calls specific functions for MIP, such as glp_mip_obj_val. One of the differences is that, for MIP problems, you are required to solve the corresponding LP problem first, as a way to create an initial feasible solution for the continuous problem. After that, you can call the MIP solver, which will create the solution-search algorithm based on a tree of possible integer values.

Complete Code

Listing 12-3 displays the complete code for the MIP solver described in the previous section. You can test the MIPSolver class using the sample code in the main function at the end of Listing 12-3.

Listing 12-3. MIPSolver Class

```
//
//  MIPSolver.h

#ifndef __FinancialSamples__MIPSolver__
#define __FinancialSamples__MIPSolver__

#include "LPSolver.h"

class MIPSolver : public LPSolver {
public:
    MIPSolver(Matrix &A, const std::vector<double> &b, const
    std::vector<double> &c);
    MIPSolver(const MIPSolver &p);
    ~MIPSolver();
    MIPSolver &operator=(const MIPSolver &p);

    void setColInteger(int colNum);
    void setColBinary(int colNum);
    virtual ResultType solve(std::vector<double> &result, double
    &objValue);
};

#endif /* defined(__FinancialSamples__MIPSolver__) */
```

```cpp
//
//  MIPSolver.cpp

#include "MIPSolver.h"

#include "Matrix.h"

#include <glpk.h>
#include <iostream>

using std::vector;
using std::cout;
using std::endl;

MIPSolver::MIPSolver(Matrix &A, const std::vector<double> &b, const
std::vector<double> &c)
: LPSolver(A, b, c)
{
}

MIPSolver::MIPSolver(const MIPSolver &p)
: LPSolver(p)
{
}

MIPSolver::~MIPSolver()
{
}

MIPSolver &MIPSolver::operator=(const MIPSolver &p)
{
    return *this;
}

void MIPSolver::setColInteger(int colNum)
{
    glp_set_col_kind(getLP(), colNum+1, GLP_IV);
}
```

```
void MIPSolver::setColBinary(int colNum)
{
    glp_set_col_kind(getLP(), colNum+1, GLP_BV);
}

LPSolver::ResultType MIPSolver::solve(vector<double> &result, double
&objValue)
{
    glp_simplex(getLP(), NULL);
    int res = glp_intopt(getLP(), NULL);
    if (res != 0)
    {
        cout << "res = " << res << " \n";
    }

    result.resize(getNumCols(), 0);
    objValue = glp_mip_obj_val(getLP());

    for (int i=0; i<getNumCols(); ++i)
    {
        result[i] = glp_mip_col_val(getLP(), i+1);
    }
    return LPSolver::FEASIBLE;
}

int main()
{
    Matrix A(2,2);
    vector<double> b = { 2, 3 };
    vector<double> c = { 1, 1 };
    A[0][0] = 1;
    A[0][1] = 2;
    A[1][0] = 3;
    A[1][1] = 4;
    MIPSolver solver(A, b, c);

    solver.setMaximization();
    solver.setColInteger(0);
```

```
    vector<double> result;
    double objVal;
    solver.solve(result, objVal);
    cout << "optimum: " << objVal << endl;
    cout << " x0: " << result[0] << " x1: " << result[1] << endl;
    return 0;
}
```

Running the Code

After compiling the code in Listing 12-3 with your favorite compiler, you should have a binary program, which I will call mipSolver. The following is the output of the application after it is executed:

./mipSolver
```
GLPK Simplex Optimizer, v4.54
2 rows, 2 columns, 4 non-zeros
*     0: obj = 0.000000000e+00 infeas = 0.000e+00 (0)
*     1: obj = 1.000000000e+00 infeas = 0.000e+00 (0)
OPTIMAL LP SOLUTION FOUND
GLPK Integer Optimizer, v4.54
2 rows, 2 columns, 4 non-zeros
1 integer variable, none of which are binary
Integer optimization begins...
+     1: mip =     not found yet <=                     +inf        (1; 0)
+     1: >>>>>   1.000000000e+00 <=   1.000000000e+00   0.0% (1; 0)
+     1: mip =   1.000000000e+00 <=       tree is empty   0.0% (0; 1)
INTEGER OPTIMAL SOLUTION FOUND
optimum: 1
 x0: 1 x1: 0
```

Notice that the main function has a very simple MIP model as part of the test code. The previous output just shows how the data is printed by default by GLPK. At each step of the process, it shows the current solution found and its objective cost.

Conclusion

Optimization is a set of mathematical techniques that can be used to find the maximum or minimum of a function under certain conditions. Since many problems in finance can be described as maximizing certain outcomes (such as investment return), optimization tools play an important role in the analysis of investments.

You have learned in this chapter how to use C++ to model and solve common optimization problems. In the first section, I showed how to use GLPK, a popular optimization library that is used to solve a large class of LP models. While other libraries such as cplex and gurobi use more sophisticated algorithms, GLPK is able to solve a surprisingly large number of models for LP and MIP. The example presented in the first section shows how to create a C++ interface to interact with the C-based API supported by GLPK. You have learned the main components of an LP problem and how the solver uses these components to determine the optimal solution. The example also provided some test code, which shows how the LPSolver class can be used to solve a simple LP model.

The next section provided a substantial example of LP model, targeted at finding the best resource allocation in a big investment bank. The model tries to find the best way to develop two financial products while maximizing the department profits. The constraints considered in the model have to do with resource limitations at the bank unit. Solving this LP problem using the LPSolver class shows how to model such problems in C++ and interpret the results of the optimization process returned by GLPK.

MIP is another important class of linear models that can be solved using optimization techniques. In an MIP, some or all decision variables are restricted to contain only integer values. This makes it possible to model situations where it is necessary to decide between two or more scenarios in a mutually exclusive way. While MIP models may be much harder to solve than LP models, the mechanics of setting up the problem and using a solver are very similar. You learned how to interface with an MIP solver by inheriting from the LPSolver class. The new class MIPSolver uses most of the modeling mechanisms provided by LPSolver, but it adds the ability to define integer and binary decision variables, as well as to solve the problem and retrieve the optimal integer values. Finally, you have seen a small example of how such models work. Chapter 13 provides more examples.

You have so far learned some of the basic concepts of optimization and mathematical programming models. In the next chapter, you will explore how these LP and MIP optimization models can be applied to investment management problems. Since one of the main applications of optimization theory is in the area of finance, many common problems such as portfolio management have well-known optimization models. You will learn how such modeling concepts work, as well as how these models can be implemented with the C++ programming language.

CHAPTER 13

Asset and Portfolio Optimization

Portfolio managers have to face several investment issues such as rebalancing a portfolio for optimal performance or adjusting a new set of investments depending on their client's predefined long-term goals. Optimization-based techniques have been developed over the years to deal with these as well as some other common portfolio construction problems.

In this chapter, you will explore programming algorithms for asset and portfolio optimization using C++ as a modeling language. You will be able to create such financial models based on well-known mathematical programming formulations. You will also see how to improve the performance of such optimization code in order to get results that are as fast and accurate as possible.

The following are some of the topics that will be covered in the C++ examples contained in this chapter:

- Allocating capital: One of the great problems faced by companies and banks is how to allocate capital to a set of possible investments. You will see how to use optimization models to perform capital allocation.

- Creating a portfolio by target return: You can use optimization models to design a portfolio of stocks or other investments, based on the desired return. The goal of such optimal portfolios is to achieve the best return with minimum volatility.

- Linear and quadratic models: You will learn the advantages of quadratic and linear optimization models for portfolio optimization.

© Carlos Oliveira 2021
C. Oliveira, *Practical C++20 Financial Programming*, https://doi.org/10.1007/978-1-4842-6834-6_13

Financial Resource Allocation

In this section, we write a linear programming (LP) model in C++ to determine an optimal allocation of resources for a given set of projects and their respective costs during a 10-year horizon.

Solution

Resource allocation is one of the most common problems faced by individual and institutional investors. Since capital is a limited resource, it makes sense to try to improve its utilization, so that one can achieve the optimal allocation of funds to valuable activities. Even though investment outcomes, such as stock prices, may not be totally predictable, it is still possible to use a general forecast for the purpose of decision making.

Linear programming offers a framework for financial allocation decisions, as you will see in this section. In the first place, you need to determine the form of the linear program that can be best used to model the resource allocation problem.

To work with a concrete allocation example, suppose that a company needs to decide among a set of five different active investments. These investments may include buying new manufacturing equipment, hiring new workers for a business, or making improvements on a logistic software package. All of these options have a specific cost, which can be calculated for each of the next 5 years. Moreover, the payoff of each investment project is known in advance. For example, if the money is invested in buying new equipment, it is known that a certain amount of profits will be generated as a result.

As a financial developer for this company, your tasks would be to implement a model to solve the required financial allocation problem. This can be done as an LP model, which will later be implemented in C++. So, first, let's consider the variables, constraints, and objective function of the LP model.

The decision variable in this case is a choice on the possible investment. That is, if there are n possible investments, then we have variables $x_j = 1$, for $j \in \{1,..,n\}$, whenever capital is allocated to project j. If the return for each investment is denoted by r_j, then we can write the objective function of this LP as

$$\max \sum_{j=1}^{n} r_j x_j.$$

The constraints are related to the amount of money that investors want to use each year for the next 5 years. Since the cost of each investment is known for any of the m periods, let's name such costs c_{ij}, for $i \in \{1...m\}$ and $j \in \{1...n\}$. For each year, the investment is limited by the value C_i, the amount of capital available at time period i. Then, for each time period (where each period corresponds to 1 year), the constraint can be written as

$$\sum_{i}^{n} c_{ij} x_j \le C_i;\, for\, i \in \{1...m\}.$$

Finally, we defined each variable x_j as a one-or-nothing decision. That is, the variable can only assume values 1 or 0, indicating that the project will be pursued or not.

$$x_j \in \{0,1\},\, for\, j \in \{1,...,n\}.$$

Because the problem described previously has a linear objective function and linear constraints, it is a linear optimization problem. However, the last constraint makes the problem a 0–1 integer LP problem, which can be considerably more difficult to solve than a standard LP problem.

Implementation

To implement the problem described previously, I will take advantage of the MIPSolver class defined in the previous chapter. Remember that the input for any mixed integer-programming problem can be represented using a matrix of constraints, a vector or right-hand side values, and a vector of costs. Thus, we need to define these three elements when defining the desired capital allocation problem.

To give a clear demonstration of how this process works, I created a simple example that can be viewed in the member function solveProblem, which is part of the class ResourceAlloc. First, this method defines a matrix of project costs for a period of 5 years. We also have five projects, so this results in a square matrix—notice, however, that a square matrix is not necessary for this formulation to work.

The next few lines of the method solveProblem define the investment returns and annual budgets. An important part of this process is to use the setBinary member function, which says that each variable must have a binary value. Finally, you need to call the function solve in the MIPSolver class, which will call the Gnu Linear Programming Kit (GLPK) solver and determine the optimum values.

Complete Code

The complete code for the resource allocation problem described in the previous section can be viewed in Listing 13-1. The main function at the end of the listing will instantiate the ResourceAlloc class and solve the example problem.

Listing 13-1. Cass ResourceAlloc

```
//
//  ResourceAlloc.h

#ifndef __FinancialSamples__ResourceAlloc__
#define __FinancialSamples__ResourceAlloc__

#include <vector>

class ResourceAlloc {
public:
    ResourceAlloc(std::vector<double> &result, double &objVal);
    ResourceAlloc(const ResourceAlloc &p);
    ~ResourceAlloc();
    ResourceAlloc &operator=(const ResourceAlloc &p);

    void solveProblem();
private:
    std::vector<double> &m_results;
    double &m_objVal;
};

#endif /* defined(__FinancialSamples__ResourceAlloc__) */

//
//  ResourceAlloc.cpp

#include "ResourceAlloc.h"

#include "LPSolver.h"
#include "Matrix.h"
```

```cpp
#include <iostream>

using std::vector;
using std::cout;
using std::endl;

ResourceAlloc::ResourceAlloc(vector<double> &result, double &objVal)
: m_results(result),
  m_objVal(objVal)
{
}

ResourceAlloc::ResourceAlloc(const ResourceAlloc &p)
: m_results(p.m_results),
  m_objVal(p.m_objVal)
{
}

ResourceAlloc::~ResourceAlloc()
{
}

ResourceAlloc &ResourceAlloc::operator=(const ResourceAlloc &p)
{
    if (this != &p)
    {
        m_results = p.m_results;
        m_objVal = p.m_objVal;
    }
    return  *this;
}

void ResourceAlloc::solveProblem()
{
    static const double cost_array[][5] = {
        // Years:
        // 1    2    3     4     5
        {1.81, 2.4,  2.5, 0.97, 1.5},  // proj 1
        {1.29, 1.8,  2.3, 0.56, 0.5},  // proj 2
```

```cpp
        {1.22, 1.2,  0.1, 0.48, 0 },    // proj 3
        {1.43, 1.4,  1.2, 1.2, 1.2},    // proj 4
        {1.62, 1.9,  2.5, 2.0, 1.8},    // proj 5
    };

    Matrix costs(5,5);  // cost matrix
    for (int i=0; i<5; ++i) {
        for (int j=0; j<5; ++j) {
            costs[j][i] = cost_array[i][j];
        }
    }

    vector<double> returns = {12.13, 3.95, 7.2, 4.21, 11.39};
    // investment returns
    vector<double> budgets = {5.1, 6.4, 6.84, 4.5, 3.8};       // annual
                                                                  budgets

    MIPSolver solver(costs, budgets, returns);
    solver.setMaximization();

    for (int i=0; i<5; ++i)
    {
        solver.setColBinary(i);
    }

    // --- solve the problem
    solver.solve(m_results, m_objVal);
}

int main()
{

    vector<double> result;
    double objVal;

    ResourceAlloc ra(result, objVal);
    ra.solveProblem();
    cout << " optimum: " << objVal ;
```

```
for (int i=0; i<result.size(); ++i)
{
    cout << " x" << i << ": " << result[i];
}
cout << endl;
return 0;
}
```

Running the Code

To run the code presented in Listing 13-1, you need to first compile it using a standards-compliant compiler such as gcc or Visual Studio. Then, you can run the resulting executable to view the results of the optimization process.

./investAllocSolver

```
GLPK Simplex Optimizer, v4.54
5 rows, 5 columns, 24 non-zeros
*      0: obj =   0.000000000e+00  infeas =  0.000e+00 (0)
*      5: obj =   3.209790698e+01  infeas =  0.000e+00 (0)
OPTIMAL LP SOLUTION FOUND
GLPK Integer Optimizer, v4.54
5 rows, 5 columns, 24 non-zeros
5 integer variables, all of which are binary
Integer optimization begins...
+      5: mip =     not found yet <=                +inf        (1; 0)
Solution found by heuristic: 30.72
+      6: mip =   3.072000000e+01 <=     tree is empty   0.0% (0; 1)
INTEGER OPTIMAL SOLUTION FOUND
 optimum: 30.72 x0: 1 x1: 0 x2: 1 x3: 0 x4: 1
Program ended with exit code: 0
```

Portfolio Optimization

Create a C++ class that can be used to define an optimal portfolio according to an LP variation of the capital asset pricing model.

Solution

One of the main uses of optimization models in finance is in the determination of investment portfolios. While there are several techniques to create balanced portfolios, the mathematical theory developed by Nobel Prize winner Harry Markowitz is the standard way to define an optimal portfolio, which is used by most financial institutions when analyzing groups of investments. In this section, you will learn about the definition of portfolio optimization models using this technique of analysis, commonly referred to as modern portfolio theory.

The main goal of portfolio optimization is to create portfolios of financial assets that can provide the required investment return with a minimum of risk. For example, if the goal is to have a small return but very low risk, one can buy high-grade investments such as US treasury bills. For higher returns, one can invest in foreign or company bonds. For even higher returns, you can use stocks and exotic derivatives.

Faced with these options, and depending on an investor profile, a portfolio manager can create one or more portfolios that address the perceived client needs. For example, a more aggressive investor may request a portfolio with a larger number of high-volatility stocks, expecting therefore a higher return. Another, more conservative investor may prefer to hold bonds and stocks with lower volatility but also lower expected returns. It is also possible to combine different portfolios to achieve a mix of high- and low-return assets.

This type of portfolio construction strategy was studied and formalized at the end of the 1950s and became known as the capital asset pricing (CAP) model. The ideas, developed by Markowitz, used classical optimization theory to characterize the optimal solutions for such portfolio construction problems. While there is a difference between finding an optimal allocation and really achieving the desired return in the financial markets, the CAP is a very important tool for portfolio managers. It can be used, for example, to define an initial portfolio that matches a particular person's profile, or to create financial products that target a defined long-term return (e.g., retirement portfolios for pension funds).

The mathematical formulation of the CAP can be summarized in the following way. Considering that there are n stocks and other assets in a portfolio, let x_i for $i \in \{1..n\}$ be the percentage of the portfolio held at investment i. Then it is clear that the sum of all such values needs to add to 1.

$$\sum_{i=1}^{n} x_i = 1$$

Also, suppose that for each investment i, we have a target return r_i (e.g., you can use past information as a baseline forecast). If the target return for the whole portfolio is R, then we have the following constraint:

$$\sum_i^n r_i x_i \geq R.$$

Now, in the CAP model, we assume that we know the variance of each asset as well as the covariance of pairs of assets in the same portfolio. Variance is a classical measure of volatility of investments (i.e., the higher the variance, the higher the volatility). Thus, we can use the available individual volatility information to try to minimize the volatility of the whole portfolio. Since variance is a quadratic function, the objective function will also be quadratic, with individual terms depending on the individual variance of individual stocks (c_{ii}) and on the covariance of pairs of stocks (c_{ij}). The resulting problem can be described as follows:

$$\min \sum_{i=1}^n \sum_{j=1}^n c_{ij}\, x_i\, x_j$$

$$\sum_{i=1}^n x_i = 1$$

$$\sum_i^n r_i x_i \geq R.$$

$$x_i \geq 0 \; for \; all \; i \in \{1\ldots n\}.$$

During the last few decades, many people have studied this optimization problem and its variations. The formulation employs an objective function that is quadratic (nonlinear)—that is, there are terms in the objective function that involve a multiplication of two variables. The general solution of this problem, considering this nonlinear structure, forms what is called an efficient frontier: a set of results for different combinations of portfolios, where the volatility of the target portfolio is minimized. You can see an example of efficient frontier in Figure 13-1, which shows a plot of volatility against target return. The plot is created by, at each time, fixing a desired return and then using the quadratic optimization model to find the minimum associated volatility. As you see in Figure 13-1, the plot shows that the relationship assumes the shape of a parabolic curve.

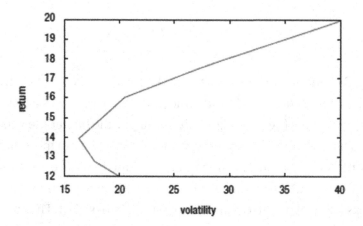

Figure 13-1. *A small portion of the efficient frontier for a portfolio optimization problem*

Although the quadratic model for CAP is widely used, a difficulty with it is the fact that you need a quadratic optimization solver to get results for a particular portfolio. Although several packages provide direct solutions to quadratic problems (using, for example, an interior point algorithm), GLPK is not able to solve quadratic optimization models directly. Therefore, in this section, you will deal with a linearization of the original problem, which can be readily calculated using LP solvers.

This linearization was proposed in Konno and Yamazaki's article titled "Mean-Absolute Deviation Portfolio Optimization Model and Its Applications to Tokyo Stock Market" (*Management Science*, vol. 37, pp. 519–529, 1991). The linearization is a modified form of the original problem that contains only linear terms in the objective function. While this is just an approximation of the original problem, in many cases it can work well enough (it might not work, however, when the computational effort needed to linearize the constraints becomes too costly). More important, a linearized version of the problem may be solved more quickly than the quadratic version, which may be an important consideration in some cases.

Consider the additional variables $y_i \in \{1...T\}$, where T is the number of periods for the proposed investment. Then, a linear model can be described as the following:

$$\min \frac{1}{T} \sum_{t=1}^{T} y_t$$

$$\sum_{j=1}^{n} \left(r_{jt} - r_j\right) x_j \leq y_t, \, for \, t \in \{1...T\}$$

$$-\sum_{j=1}^{n} \left(r_{jt} - r_j\right) x_j \leq y_t, \, for \, t \in \{1...T\}$$

$$\sum_{i=1}^{n} x_i = 1$$

$$\sum_{i}^{n} r_i x_i \geq R.$$

$$x_i \geq 0 \, for \, all \, i \in \{1...n\}$$

In these equations, you don't need directly the covariance c_{ij}; instead, you use the expected returns r_{it} or investment i for period t. In other words, the idea of the model is to divide the total periods into small segments and linearize the model during that small period, taking the minimum over the complete time horizon.

With the model, which is linear, you can now create code in C++ using the LPSolver class described in the previous chapter. The new class is called ModifiedCAP and is displayed in the next section, "The Code." The main difficulty in creating the model is defining the required input data for LPSolver, in the form of matrix A and vectors b and c. You can see how this is done in the code for member function solveModel.

The first part of the algorithm consists of setting up the required data. The vector c that defines the objective function can be easily created, since all coefficients are equal to 1.

```
// objective function
for (int i=m_N; i<m_N+m_T; ++i)
{
    c[i] = 1;
}
```

Next, the right-hand side coefficients are also simple to set up. This is true because you can move all variables y_t to the left side of the inequality. Thus, most of the coefficients are zero, except for the last three.

```
// right-hand side vector
vector<double> b(2*m_T + 2 + 1, 0);
b[2*m_T]   =  1;
b[2*m_T+1] = -1;
b[2*m_T+2] = -m_R;
```

Matrix A is a little more involved but is not difficult to set up either. The main transformation you need to make is in the equality constraints. Since the problems that LPSolver considers have inequalities only, the equality $\sum_{i=1}^{n} x_i = 1$ is handled by converting it into two inequalities.

$$\sum_{i=1}^{n} x_i \leq 1 \ and - \sum_{i=1}^{n} x_i \leq -1$$

This makes it possible to continue to use the same simple input form used by the LPSolver (GLPK can also handle equalities directly, and you could modify the LPSolver class to do this automatically). Therefore, the following code can be used to define the input matrix A:

```
// matrix A
Matrix A(2*m_T + 2 + 1, m_T + m_N);
for (int i=0; i<m_T; ++i)
{
    for (int j=0; j<m_N; ++j)
    {
        A[i][j] = m_retMatrix[j][i] - m_assetRet[j];
    }
}
```

```
    A[i][m_N+i] = -1;
}

for (int i=m_T; i<2*m_T; ++i)
{
    for (int j=m_N; j<2*m_N; ++j)
    {
        A[i][j] = - m_retMatrix[j-m_N][i-m_T] + m_assetRet[j-m_N];
    }
    A[i][m_N+i-m_T] = -1;
}

for (int j=0; j<m_N; ++j)
{
    A[2*m_T][j]   = 1;
    A[2*m_T+1][j] = -1;
    A[2*m_T+2][j] = - m_assetRet[j];
}
```

The remainder of the code is just to handle constructing the LPSolver class and calling the required member functions to solve the model.

Finally, I provide a simple example of how this class could be called in practice. The sample data has four assets and five time periods. The associated expected returns are given by the following matrix, which you will find in the test main function:

```
// sample problem: 4 assets and 5 periods

// build the expected return matrix
double val[][5] = {
    {0.051, 0.050, 0.049, 0.051, 0.05},
    {0.10, 0.099, 0.102, 0.100, 0.101},
    {0.073, 0.077, 0.076, 0.075, 0.076},
    {0.061, 0.06, 0.059, 0.061, 0.062},
};
```

Complete Code

You can view the complete code for the modified CAP in Listing 13-2. The listing contains a header and an implementation file.

Listing 13-2. Modified CAP Implementation

```cpp
//
//  ModifiedCAP.h

#ifndef __FinancialSamples__ModifiedCAP__
#define __FinancialSamples__ModifiedCAP__

#include "Matrix.h"

// a modified (linearized) model for Capital Asset Pricing
class ModifiedCAP {
public:
    ModifiedCAP(int N, int T, double R, Matrix &retMatrix, const
    std::vector<double> &ret);
    ModifiedCAP(const ModifiedCAP &p);
    ~ModifiedCAP();
    ModifiedCAP &operator=(const ModifiedCAP &p);

    void solveModel(std::vector<double> &results, double &objVal);
private:
    int m_N;  // number of investment
    int m_T;  // number of periods
    double m_R;  // desired return
    Matrix m_retMatrix;
    std::vector<double> m_assetRet; // single returns
};

#endif /* defined(__FinancialSamples__ModifiedCAP__) */

//
//  ModifiedCAP.cpp

#include "ModifiedCAP.h"
```

```cpp
#include "LPSolver.h"

#include <iostream>
#include <vector>

using std::vector;
using std::cout;
using std::endl;

ModifiedCAP::ModifiedCAP(int N, int T, double R, Matrix &expectedRet, const
vector<double> &ret)
: m_N(N),
  m_T(T),
  m_R(R),
  m_retMatrix(expectedRet),
  m_assetRet(ret)
{
}

ModifiedCAP::ModifiedCAP(const ModifiedCAP &p)
: m_N(p.m_N),
  m_T(p.m_T),
  m_R(p.m_R),
  m_retMatrix(p.m_retMatrix),
  m_assetRet(p.m_assetRet)
{
}

ModifiedCAP::~ModifiedCAP()
{
}

ModifiedCAP &ModifiedCAP::operator=(const ModifiedCAP &p)
{
    if (this != &p)
    {
        m_N = p.m_N;
        m_T = p.m_T;
```

```
        m_R = p.m_R;
        m_retMatrix = p.m_retMatrix;
        m_assetRet = p.m_assetRet;
    }
    return *this;
}

void ModifiedCAP::solveModel(std::vector<double> &results, double &objVal)
{
    Matrix A(2*m_T + 2 + 1, m_T + m_N);
    vector<double> c(m_T + m_N, 0);

    // objective function
    for (int i=m_N; i<m_N+m_T; ++i)
    {
        c[i] = 1;
    }

    // right-hand side vector
    vector<double> b(2*m_T + 2 + 1, 0);
    b[2*m_T]   =  1;
    b[2*m_T+1] = -1;
    b[2*m_T+2] = -m_R;

    // matrix A
    for (int i=0; i<m_T; ++i)
    {
        for (int j=0; j<m_N; ++j)
        {
            A[i][j] = m_retMatrix[j][i] - m_assetRet[j];
        }
        A[i][m_N+i] = -1;
    }

    for (int i=m_T; i<2*m_T; ++i)
    {
        for (int j=m_N; j<2*m_N; ++j)
```

```
        {
            A[i][j] = - m_retMatrix[j-m_N][i-m_T] + m_assetRet[j-m_N];
        }
        A[i][m_N+i-m_T] = -1;
    }

    for (int j=0; j<m_N; ++j)
    {
        A[2*m_T][j]   = 1;
        A[2*m_T+1][j] = -1;
        A[2*m_T+2][j] = - m_assetRet[j];
    }

    LPSolver solver(A, b, c);
    solver.setMinimization();
    solver.solve(results, objVal);
}

int main()
{

    // sample problem: 4 assets and 5 periods

    // build the expected return matrix
    double val[][5] = {
        {0.051, 0.050, 0.049, 0.051, 0.05},
        {0.10, 0.099, 0.102, 0.100, 0.101},
        {0.073, 0.077, 0.076, 0.075, 0.076},
        {0.061, 0.06, 0.059, 0.061, 0.062},
    };
    Matrix retMatrix(4, 5);
    for (int i=0; i<4; ++i)
    {
        for (int j=0; j<5; ++j)
        {
            retMatrix[i][j] = val[i][j];
        }
    }
```

```
vector<double> assetReturns = {0.05, 0.10, 0.075, 0.06};

ModifiedCAP mc(4, 5, 0.08, retMatrix, assetReturns);

vector<double> results;
double objVal;
mc.solveModel(results, objVal);

cout << "obj value: " << objVal/5 << endl;
for (int i=0; i<results.size(); ++i)
{
    cout << " x" << i << ": " << results[i];
}
cout << endl;
}
```

Running the Code

The ModifiedCAP class presented in Listing 13-2 can be compiled using any standards-compliant C++ compiler. The main class depends on other classes presented before, such as LPSolver and Matrix. The code also depends on the GLPK library, which you can download for free as described in the previous chapter. After building this class into the executable ModifiedCap, you can run the test main function and see results similar to what is shown in the following code:

```
GLPK Simplex Optimizer, v4.54
13 rows, 9 columns, 46 non-zeros
      0: obj =   0.000000000e+00  infeas =  1.080e+00 (0)
*     8: obj =   3.380952381e-03  infeas =  0.000e+00 (0)
*    10: obj =   1.881151309e-03  infeas =  1.110e-16 (0)
OPTIMAL LP SOLUTION FOUND
obj value: 0.00037623
 x0: 0.320288 x1: 0.520288 x2: 0.159424 x3: 0 x4: 1.43914e-06 x5: 0 x6:
0.000879712 x7: 0.000320288 x8: 0.000679712
Program ended with exit code: 0
```

Looking at the output, the result shows nine LP variables. From the formulation, you will see that the first four variables correspond to the original CAP variables, while the last five are related to the time periods and therefore not used in the portfolio construction. These results tell the portfolio manager that only the first three assets should be considered in the portfolio, with percentages equal to 32%, 52%, and 15%, respectively.

To improve these results, you can modify the model accordingly to your goals. For example, you can try a different return and see how the portfolio will change based on the additional information.

Extensions to Modified CAP

In this section, we create extensions to the modified CAP model so that no asset is assigned more than 30% of the portfolio. Also, add a rule that asset classes gold and treasury bills compose at least 15% of the portfolio.

Solution

In the section "Portfolio Optimization," you saw how to create an optimization model to determine the optimal allocation of capital to a specified portfolio so that the required target return is achieved while minimizing the volatility of the resulting portfolio. The given formulation is a modification of the original method proposed in CAP, which is a quadratic optimization model. Despite this, you can achieve quite fast results using a linear programming version of the model.

Although this model is able to cover the basis of a portfolio construction strategy, you can try other useful variations. For example, a common modification of the LP model presented previously consists of adding the minimum and maximum requirement for each asset type.

For example, suppose that you may want to increase the diversification of your portfolio by enforcing a limit on the percentage held of each asset. The main idea here is to avoid big losses that result from a portfolio concentrated in a small number of assets. Such a requirement could be easily added to the model with the following constraint:

$$x_j \leq M, \text{ for each } j \in \{1...n\}$$

Here, M is the desired percentage limit. When run, the LP solver will guarantee using this constraint that each percentage is not greater than the given amount M.

Similarly, you can also define a minimum amount held for each asset. In this case, it is frequently useful to have separate minimum values for each possible investment. For example, you may want to have a portfolio where treasury bills will be at least 5% at any time. If we denote the minimum required allocation by K_j, this would lead to a constraint of the type

$$x_j \geq K_j, \text{ for each } j \in \{1...n\}$$

In general, similar modifications can also be done for combinations of assets, such as treasury bills and gold. This would also work for larger groups of assets, such as adding a minimum threshold for the total number of all growth stocks in a portfolio. If you have a group of stocks L and an associated limit K_L, then this general constraint can be denoted by

$$\sum_{j \in L} x_j \geq K_L.$$

Finally, you can also use the idea of groups of assets to define an upper bound of the percentage held in these investments. For example, if you want to limit the percentage of a portfolio exposed to technology stocks, you can denote the group by U and the limit by K_U, resulting in the following constraint:

$$\sum_{j \in U} x_j \leq K_U.$$

For the benefit of simplicity, I have provided an alternative version of the `ModifiedCAP` class, where we have an alternative rule for diversification (at 37% level) and a minimum of 15% for the combined assets 1 and 2 (gold and treasury bills). The new code is implemented in the function `solveExtendedModel`, defined as

```
void solveExtendedModel(std::vector<double> &results, double &objVal);
```

The remaining parts of the class remain unchanged in this coding example.

Complete Code

You can find the code to solve the extended version of the CAP model in Listing 13-3. In the header file, I show the complete class declaration, which is similar to the previous listing except for the added function, solveExtendedModel. The implementation file shows only the new member function, along with a test main function.

Listing 13-3. Extended Model for the CAP

```
//
//  ModifiedCAP.h

#ifndef __FinancialSamples__ModifiedCAP__
#define __FinancialSamples__ModifiedCAP__

#include "Matrix.h"

// a modified (linearized) model for Capital Asset Pricing
class ModifiedCAP {
public:
    ModifiedCAP(int N, int T, double R, Matrix &retMatrix, const
    std::vector<double> &ret);
    ModifiedCAP(const ModifiedCAP &p);
    ~ModifiedCAP();
    ModifiedCAP &operator=(const ModifiedCAP &p);

    void solveModel(std::vector<double> &results, double &objVal);
    void solveExtendedModel(std::vector<double> &results, double &objVal);
private:
    int m_N;    // number of investment
    int m_T;    // number of periods
    double m_R; // desired return
    Matrix m_retMatrix;
    std::vector<double> m_assetRet; // single returns
};

#endif /* defined(__FinancialSamples__ModifiedCAP__) */

//
//  ModifiedCAP.cpp

#include "ModifiedCAP.h"

#include "LPSolver.h"
```

```cpp
#include <iostream>
#include <vector>

//
// ... just like code list displayed on previous section
//

void ModifiedCAP::solveExtendedModel(std::vector<double> &results, double
&objVal)
{
    vector<double> c(m_T + m_N, 0);

    // objective function
    for (int i=m_N; i<m_N+m_T; ++i)
    {
        c[i] = 1;
    }

    const double M = 0.37;      // maximum of each asset
    const double K_L = 0.15;    // minimum of combined assets 1 and 2

    // right-hand side vector
    vector<double> b(2*m_T + 2 + 1 + m_N + 1 , 0);
    b[2*m_T]   =  1;
    b[2*m_T+1] = -1;
    b[2*m_T+2] = -m_R;

    for (int i=2*m_T+3; i<2*m_T + 3 + m_N; ++i)
    {
        b[i] = M;
    }
    b[2*m_T + 3 + m_N] = -K_L;

    // matrix A
    Matrix A(2*m_T + 2 + 1 + m_N + 1, m_T + m_N);
    for (int i=0; i<m_T; ++i)
    {
```

```
    for (int j=0; j<m_N; ++j)
    {
        A[i][j] = m_retMatrix[j][i] - m_assetRet[j];
    }
    A[i][m_N+i] = -1;
}

for (int i=m_T; i<2*m_T; ++i)
{
    for (int j=m_N; j<2*m_N; ++j)
    {
        A[i][j] = - m_retMatrix[j-m_N][i-m_T] + m_assetRet[j-m_N];
    }
    A[i][m_N+i-m_T] = -1;
}

for (int j=0; j<m_N; ++j)
{
    A[2*m_T][j]   = 1;
    A[2*m_T+1][j] = -1;
    A[2*m_T+2][j] = - m_assetRet[j];
}

// constraints for percentage limit
for (int i=2*m_T+3; i<2*m_T+3+m_N; ++i)
{
    A[i][i-(2*m_T+3)] = 1;
}
// constraints for assets 1 and 2
A[2*m_T+3+m_N][0] = -1;
A[2*m_T+3+m_N][1] = -1;

LPSolver solver(A, b, c);
solver.setMinimization();
solver.solve(results, objVal);
}
```

```
int main()
{

    // sample problem: 4 assets and 5 periods

    // build the expected return matrix
    double val[][5] = {
        {0.051, 0.050, 0.049, 0.051, 0.05},
        {0.10, 0.099, 0.102, 0.100, 0.101},
        {0.073, 0.077, 0.076, 0.075, 0.076},
        {0.061, 0.06, 0.059, 0.061, 0.062},
    };
    Matrix retMatrix(4, 5);
    for (int i=0; i<4; ++i)
    {
        for (int j=0; j<5; ++j)
        {
            retMatrix[i][j] = val[i][j];
        }
    }

    vector<double> assetReturns = {0.05, 0.10, 0.075, 0.06};

    ModifiedCAP mc(4, 5, 0.08, retMatrix, assetReturns);

    vector<double> results;
    double objVal;
    mc.solveExtendedModel(results, objVal);

    cout << "obj value: " << objVal/5 << endl;
    for (int i=0; i<results.size(); ++i)
    {
        cout << " x" << i << ": " << results[i];
    }
    cout << endl;
    return 0;
}
```

Running the Code

After compiling the class described in Listing 13-3, you will be able to find the modified results of the optimized portfolio. The following is a sample output of what I achieved by adding the constraints explained previously:

```
./extendedModifiedCAP
GLPK Simplex Optimizer, v4.54
18 rows, 9 columns, 52 non-zeros
      0: obj = 0.000000000e+00 infeas = 1.230e+00 (0)
*    14: obj = 2.671440000e-03 infeas = 0.000e+00 (0)
OPTIMAL LP SOLUTION FOUND
obj value: 0.000534288
 x0: 0.035 x1: 0.37 x2: 0.37 x3: 0.225 x4: 1.44e-06 x5: 0.00037 x6: 0.00085
x7: 0.00026 x8: 0.00119
Program ended with exit code: 0
```

The solution found by the optimizer shows that the optimum allocation for the four asset classes would be 3.5%, 37%, 37%, and 22%, respectively.

Conclusion

Portfolio optimization is a tool frequently used by portfolio managers to help define a suitable capital allocation depending on the desired goals of their clients. Therefore, it is important for financial C++ programmers to be able to devise efficient solutions for such portfolio allocation problems.

In this chapter, you have learned a few mathematical programming models that have been successfully used by financial institutions to create and manage portfolios as well as other financial allocation problems. In the first section, you have seen how mixed-integer programming (MIP) can be used to model some financial allocation problems. You learned about some of the differences between MIP and LP models and how they can be solved with the help of the `MIPSolver` class.

In the next section, the focus was on the CAP model, where the main goal is to determine the percentages of each investment that need to be held in a portfolio, in order to achieve the desired outcome while at the same time minimizing the associated volatility of the investments. You have seen that although this is a quadratic programming problem, it is possible to achieve good results with a linearization of the mathematical formulation. An alternative formulation was presented and implemented in C++ using the `AlternativeCAP` class.

Finally, I discussed a few extensions of the basic model and how the optimization results can be understood in the context of the required portfolio. Such extensions to the basic CAP model are common, and they help in developing portfolios that are subject to real-world constraints for asset allocation.

In the next chapter, you will become acquainted with another key technique in financial engineering: Monte Carlo simulation methods. You will see a few C++ examples of how such techniques can be quickly implemented and how the results of such methods can be interpreted.

CHAPTER 14

Monte Carlo Methods

Among other programming techniques for equity markets, Monte Carlo simulation has a special place due to its wide applicability and relatively easy implementation compared to exact, non-stochastic methods. These algorithms can be used in many applications such as price forecasting and the validation of certain buying strategies, for example.

In this chapter, we provide C++ programming code that can be used either directly or as part of simulation-based algorithms. These examples will introduce some of the most important concepts used in the development of stochastic methods. The following is a quick summary of topics discussed in this chapter:

- Determining definite integrals: Random sampling is a powerful way to calculate complicated functions with a minimum of computational effort. You will see how to use stochastic techniques to determine definite integrals.

- Forecasting prices: Being a common technique to simulate random price fluctuations, Monte Carlo methods have been frequently used as a way to forecast prices. The ability to repeat the simulation process is a key feature of this method.

- Calculating options prices: Among other methods for option pricing forecasting, Monte Carlo techniques have been widely used due to its simplicity. Unlike other mathematical methods, simulations can be quickly coded and generally perform well compared to exact techniques for option price forecasting.

Monte Carlo-Based Integral Computation

Create a class to estimate the integral of a generic function using a Monte Carlo strategy.

377

© Carlos Oliveira 2021
C. Oliveira, *Practical C++20 Financial Programming*, https://doi.org/10.1007/978-1-4842-6834-6_14

Solution

The main concept of Monte Carlo methods is to use a random process to find solutions for a complex problem. While a random solution may not be useful for the problem at hand, it has the important property that it can be repeated with different results. The information that you can gather by looking at a large number of such Monte Carlo results is the secret of such techniques.

A classic example of using Monte Carlo methods is determining the area defined by a curve with random sampling. For example, to find the area of a circle, you can draw several random points and check if they are part of the circle. The area is then determined by the percentage of points inside the circle. As the number of points increase, you will get better approximations for the required area.

An extension of the general idea described previously is the basis for a Monte Carlo strategy for integration. The advantage of using such Monte Carlo methods to integrate functions is that you just need to generate random points in the given range. The simplicity of the strategy makes it possible to estimate the integral of very complicated functions with a minimum of code.

You can see an implementation of this method in the MonteCarloIntegration class. The structure of the class is similar to the examples you saw in Chapter 10, which covers integration. However, the algorithm used involves the generation of random samples, in order to determine the percentage of area under the given function.

To generate uniformly distributed random numbers, we use the uniform_real_ distribution class, part of boost::random. This simplifies the generation of samples, avoiding numerical accuracy issues that are common when using other sources of random numbers.

The main part of the implementation can be viewed in the getIntegral member function.

```
double MonteCarloIntegration::getIntegral(double a, double b)
```

The code initially determines the maximum and minimum values observed. It uses these numbers to define the total area of sampling. Then, the function generates random numbers and checks if they are inside the curve defined by the function or outside it. At the end, the percentage calculated with this procedure is used to compute the total area of the integral. This process is repeated for the positive and negative parts of the given mathematical function, using the member function integrateRegion. The total value of the integral is then calculated as the positive minus the negative area.

Complete Code

You will find the complete code to integrate a function using the Monte Carlo methods in Listing 14-1. The code is divided into a header and an implementation file. A sample main function is included to show how the class MonteCarloIntegration can be used.

Listing 14-1. Monte Carlo Integration Method

```
//
// MonteCarloIntegration.h

#ifndef __FinancialSamples__MONTECARLOINTEGRATION_H_
#define __FinancialSamples__MONTECARLOINTEGRATION_H_

template <class T>
class MathFunction;

class MonteCarloIntegration {
public:
    MonteCarloIntegration(MathFunction<double> &f);
    MonteCarloIntegration(MathFunction<double> &f, int num_samples);
    MonteCarloIntegration(const MonteCarloIntegration &p);
    ~MonteCarloIntegration();
    MonteCarloIntegration &operator=(const MonteCarloIntegration &p);

    void setNumSamples(int n);
    double getIntegral(double a, double b);
    double integrateRegion(double a, double b, double min, double max);
private:
    MathFunction<double> &m_f;
    int m_numSamples;

};

#endif /* MONTECARLOINTEGRATION_H_ */

//
// MonteCarloIntegration.cpp

#include "MonteCarloIntegration.h"
```

```cpp
#include <cmath>
#include <cstdlib>
#include <iostream>

#include "MathFunction.h"

#include <random>

static std::default_random_engine random_generator;

using std::cout;
using std::endl;

namespace {
    const int DEFAULT_NUM_SAMPLES = 1000;
}

MonteCarloIntegration::MonteCarloIntegration(MathFunction<double>& f)
: m_f(f),
  m_numSamples(DEFAULT_NUM_SAMPLES)
{
}

MonteCarloIntegration::MonteCarloIntegration(MathFunction<double>& f, int
num_samples)
: m_f(f),
  m_numSamples(num_samples)
{
}

MonteCarloIntegration::MonteCarloIntegration(const MonteCarloIntegration&
p)
: m_f(p.m_f),
  m_numSamples(p.m_numSamples)
{
}

MonteCarloIntegration::~MonteCarloIntegration()
{
}
```

```
MonteCarloIntegration& MonteCarloIntegration::operator =(const
MonteCarloIntegration& p)
{
    if (this != &p)
    {
        m_f = p.m_f;
        m_numSamples = p.m_numSamples;
    }
    return *this;
}

void MonteCarloIntegration::setNumSamples(int n)
{
    m_numSamples = n;
}

double MonteCarloIntegration::integrateRegion(double a, double b, double
min, double max)
{
    std::uniform_real_distribution<> xDistrib(a, b);
    std::uniform_real_distribution<> yDistrib(min, max);

    int pointsIn = 0;
    int pointsOut = 0;
    bool positive = max > 0;

    for (int i = 0; i < m_numSamples; ++i)
    {
        double x = xDistrib(random_generator);
        double y = m_f(x);

        double ry = yDistrib(random_generator);
        if (positive && min <= ry && ry <= y)
        {
            pointsIn++;
        }
        else if (!positive && y <= ry && ry <= max)
        {
```

```cpp
                pointsIn++;
        }
        else
        {
                pointsOut++;
        }
    }

    double percentageArea = 0;
    if (pointsIn+pointsOut > 0)
    {
        percentageArea = pointsIn / double(pointsIn + pointsOut);
    }

    if (percentageArea > 0)
    {
        return (b-a) * (max-min) * percentageArea;
    }

    return 0;
}

double MonteCarloIntegration::getIntegral(double a, double b)
{
    std::uniform_real_distribution<> distrib(a, b);

    double max = 0;
    double min = 0;

    for (int i = 0; i < m_numSamples; ++i)
    {
        double x = distrib(random_generator);
        double y = m_f(x);

        if (y > max)
        {
            max = y;
        }
        if (y < min)
```

```
        {
            min = y;
        }
    }
    double positiveIntg = max > 0 ? integrateRegion(a, b, 0, max) : 0;
    double negativeIntg = min < 0 ? integrateRegion(a, b, min, 0) : 0;
    return positiveIntg - negativeIntg;
}

// Example function

namespace  {

    class FSin : public MathFunction<double>
    {
    public:
        ~FSin();
        double operator()(double x);
    };

    FSin::~FSin()
    {
    }

    double FSin::operator()(double x)
    {
        return sin(x);
    }

}

int main()
{
    cout << "starting" << endl;
    FSin f;
    MonteCarloIntegration mci(f);
    double integral = mci.getIntegral(0.5, 4.9);
```

```
    cout << " the integral of the function is " << integral << endl;

     mci.setNumSamples(200000);
    integral = mci.getIntegral(0.5, 4.9);
    cout << " the integral of the function with 20000 intervals is
    " << integral << endl;
    return 0;
}
```

Running the Code

You can compile the files presented in Listing 14-1 using gcc or any other standards-compliant C++ compiler. The result for the sample code in the main function is the following, assuming that you named the executable as monteCarloIntegration:

```
$ ./monteCarloIntegration
 the integral of the function is 1.74
 the integral of the function with 20000 intervals is 1.6702
```

Notice that this can change depending on the random source used by the implementation. However, the values should approach the correct value as the number of samples used by the Monte Carlo method increase.

Simulating Asset Prices

Create a C++ class to mimic the price fluctuations of equities in the stock market using a random walk simulation process.

Solution

If there is an area in investment where it is difficult to find closed solutions, that area is financial forecasting. Although there are well-known economic models, any complex system such as the stock market is subject to wild fluctuations that result from so many factors, including wars, natural disasters, and personal choices of important players, among others. Due to the big difficulty of estimating such disparate events, a large part of market forecasting models assume that some form of random process is the source of price fluctuations. In this scenario, Monte Carlo techniques prove to be very useful in the simulation of future market conditions.

In this section, I present a very simple Monte Carlo model that can be used for forecasting purposes. To start the presentation, I introduce a first version of this method using a set of very simple simulation rules. Then, you will see a more complex version of this same principle, using a Gaussian distribution, in the next C++ coding example.

The basic strategy used in price forecasting is to simulate price movements using a "random walk." A random walk process is a stochastic technique in which the next state of the system is defined only by its previous state (a known price) and the probability distribution for the next possible moves. In the example presented in this C++ class, there are three next states, each of them having the same probability. As a result, at each moment in time, the price can go up, go down, or stay flat.

To simplify the example of random walk given in this section, we will assume that the prices of the underlying asset are moving according to a uniform distribution. In other words, price changes are generated in such a way that the average jump is received as a parameter. Also, the increase or decrease in price is defined using a uniformly distributed random variable, with values determined by the known average.

The code necessary to create this simulation is encapsulated in the RandomWalk class. The class is needed to store the information about parameters for the process: among these parameters are the number of steps (samples) in the Monte Carlo simulation, the initial asset price, and the average step used in the process.

Using these parameters, the getWalk member function runs the simulation and returns a vector of prices generated using this strategy. Figure 14-1 displays a sample result of this random process. Once you store the prices generated by getWalk, your code can perform additional processing, as needed. A common example where this may be useful is during the test of new trading strategies and for determination of their profitability.

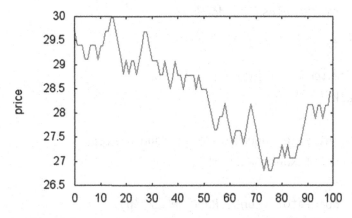

Figure 14-1. *Results of the random walk generated by RandomWalk class*

The algorithm for random walk used in the `RandomWalk` class can be tweaked in a number of ways, depending on the demands of your simulation.

- For example, you may want to have the price of the underlying instrument changing more frequently. This can be achieved with the removal of the third branching rule (which allows the price to stay in the same level) and therefore forcing moves either up or down.

- Another variation of the random walk is to have different probabilities for up and high prices—in this way, it is possible to simulate a "bull" or "bear" market.

- For a similar purpose as previously, it is possible to change the amount of the price jump (up or down), so that up moves may be bigger than down moves. This would be another way to simulate a directional market, where prices are going up faster than usual.

Complete Code

Listing 14-2 contains the complete code for the `RandomWalk` class. You will find the implementation in a header file and a cpp file, followed by a sample `main` function.

Listing 14-2. Implementation for RandomWalk Class

```
//
// RandonWalk.h

#ifndef __FinancialSamples__RandonWalk__
#define __FinancialSamples__RandonWalk__

#include <vector>

// Simple random walk for price simulation
class RandomWalk {
public:
    RandomWalk(int size, double start, double step);
    RandomWalk(const RandomWalk &p);
    ~RandomWalk();
    RandomWalk &operator=(const RandomWalk &p);
```

```
    std::vector<double> getWalk();
private:
    int m_size;     // number of steps
    double m_step;  // size of each step (in percentage)
    double m_start; // starting price
};

#endif /* defined(__FinancialSamples__RandonWalk__) */

//
//  RandonWalk.cpp

#include "RandonWalk.h"

#include <iostream>

using std::vector;
using std::cout;
using std::endl;

RandomWalk::RandomWalk(int size, double start, double step)
: m_size(size),
  m_step(step),
  m_start(start)
{
}

RandomWalk::RandomWalk(const RandomWalk &p)
: m_size(p.m_size),
  m_step(p.m_step),
  m_start(p.m_start)
{
}

RandomWalk::~RandomWalk()
{
}

RandomWalk &RandomWalk::operator=(const RandomWalk &p)
```

```cpp
{
    if (this != &p)
    {
        m_size = p.m_size;
        m_step = p.m_step;
        m_start = p.m_start;
    }
    return *this;
}

std::vector<double> RandomWalk::getWalk()
{
    vector<double> walk;
    double prev = m_start;

    for (int i=0; i<m_size; ++i)
    {
        int r = rand() % 3;
        double val = prev;
        if (r == 0) val += (m_step * val);
        else if (r == 1) val -= (m_step * val);
        walk.push_back(val);
        prev = val;
    }
    return walk;
}

int main()
{
    RandomWalk rw(100, 30, 0.01);
    vector<double> walk = rw.getWalk();
    for (int i=0; i<walk.size(); ++i)
    {
        cout << ", " << i << ", " << walk[i];
    }
    cout << endl;
    return 0;
}
```

Running the Code

To run the code presented in Listing 14-2, first you need to compile it using a standards-compliant compiler such as gcc or Visual Studio. Here are the first few lines of the random walk using the given parameters: initial price of $30, step of 1%, and 100 steps.

```
./randomWalk
, 0, 29.7, 1, 29.403, 2, 29.403, 3, 29.403, 4, 29.109, 5, 29.109, 6,
29.4001, 7, 29.4001, 8, 29.4001, 9, 29.1061, 10, 29.3971, 11, 29.3971, 12,
29.6911, 13, 29.6911, 14, 29.988, 15, 29.988, 16, 29.6881, 17, 29.3912, 18,
29.0973, 19, 28.8064, 20, 29.0944, 21, 28.8035, 22, 29.0915, 23, 29.0915,
24, 28.8006, 25, 29.0886, 26, 29.3795, 27, 29.6733, 28, 29.6733, 29,
29.3765, 30, 29.0828, 31, 29.0828, 32, 29.0828, 33, 28.792, 34, 28.792, 35,
29.0799, 36, 28.7891, 37, 28.5012, 38, 28.7862, 39, 29.0741, 40, 28.7833,
41, 28.7833, 42, 28.4955, 43, 28.7804, 44, 28.7804, 45, 28.7804, 46,
28.7804, 47, 28.4926, 48, 28.7776, 49, 28.4898, 50, 28.4898, 51, 28.4898,
52, 28.2049, 53, 27.9228, 54, 27.6436, 55, 27.6436, 56, 27.92, 57, 27.92,
58, 28.1992, 59, 27.9173,
```

Figure 14-1 displays a plot of the random walk generated by one execution of the sample code. Notice how prices start near the $30 mark and display a behavior similar to real variations observed on the stock market.

Calculating Option Probabilities

In this section, we will implement a C++ solution to computing European options probabilities for events such as finishing above the strike price, finishing below the strike price, or finishing between two given prices.

Solution

Options are a very popular type of equity derivatives, which can be bought in most retail investment accounts. With options, you pay a price for the privilege of buying or selling a stock for a particular price during a limited period of time, hence the designation "option," since you have the option, not the obligation, of performing the transaction.

A call option gives the right to buy at a particular price, while a put option gives the right to sell at a particular price. The exercise price is called the strike. Depending on the relationship between the current price of the stock and the strike price, an option can be classified into one of the three categories:

- In the money (ITM): The strike price is lower than the current price of the stock, for call options. For put options, the strike price should be above the current stock price.

- Out of the money (OTM): The strike price is higher than the current price of the stock, for call options. For put options, the strike price should be below the current stock price.

- At the money (ATM): The strike price is close to the current price of the underlying stock.

These different relationships between strike price and stock price determine different probabilities of an option to become profitable, as we will see in the remainder of this section.

To achieve profitability, a call option needs the stock price to rise above the strike price. When this happens, the price for the position is given by the difference between the strike price and the stock price, plus whatever time value the option might still have. For put options this is reversed, and the option becomes profitable when the stock prices decrease in comparison to the strike price.

Another concept in options is the style of exercise (i.e., buying or selling the underlying stock). European-style options allow the exercising of the option only at the end of its target period. American-style options, on the other hand, allow exercising to happen at any moment in time. In this section I consider European options only, since the analysis considers only the price of the stock at the option expiration. It is not hard, however, to extend the techniques explained to handle American-style options.

Figure 14-2 shows the profit profile for an option contract. The data assumes that the contract costs $5, with a strike price of $90. In this case, for any final price of the underlying stock that is below $90, the full loss of the option price is realized. On the other hand, the loss is capped, and the investor will not suffer any losses other than the price of the contract. When the underlying asset price achieves the strike price of $90, the loss of the position starts to decrease, getting to an even point at $95. From that point on, any additional price increase represents additional profit for the option position, and gains are unlimited to the upside.

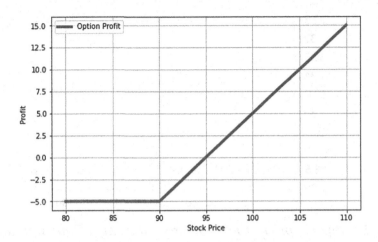

Figure 14-2. *Profit potential for an option call contract*

Determining Profit Probabilities

In this section, you will learn how to use Monte Carlo techniques to determine profit probabilities for equity options. As you have seen previously, all that is necessary to find the profit for a call option is to calculate the price of the underlying asset at expiration and check if the final price is above the strike price. For put options, the process is the same, but instead you need to check if the final price is below the strike price.

The first step in creating a Monte Carlo simulation for this problem is to define the parameters of the random process. The pricing of options is defined by what is called the Black-Scholes model, where prices are assumed to be normally distributed. For this reason, you will use a random walk with Gaussian distribution for price changes. At each step, only two possibilities are available in this random process: prices either go up or go down with 50% of chance. The price change is then determined by the normal distribution with variance that is given as an input parameter. Figure 14-3 presents an example of Gaussian random walk. Notice how similar this looks to actual price fluctuations, when compared to real data for stocks.

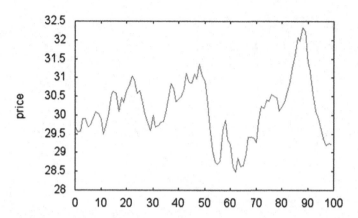

Figure 14-3. *Example of price movement created using a Gaussian random walk*

Once the random walk is generated, one can start to use the data for forecasting and related price analysis. In this case, you would like to estimate the probabilities of some events, such as finishing above a certain price level. To answer these questions, you just need to use the standard Monte Carlo procedure: repeat the random walk and store the results. After this process is performed several times, it is possible to analyze the distribution of results in which the final price was above a certain target.

For example, suppose you want to answer the question: what is the probability of the price finishing above the strike price? To do this, perform the random walk for a large number of tests, and calculate the percentage of these tests in which the price finished above the strike. The same approach can be used to collect related information, such as the probability of finishing below the strike price, or the probability of finishing between two given prices.

The implementation is given in the OptionsProbabilities class. The important member functions are the following:

```
double probFinishAboveStrike();
double probFinishBelowStrike();
double probFinalPriceBetweenPrices(double lowPrice, double highPrice);
std::vector<double> getWalk();
```

The three first member functions calculate the desired probabilities. The last member function, getWalk, returns a vector that stores a sample random walk for further analysis. The OptionsProbabilities class internally uses the getLastPriceOfWalk member function, which returns the last observed price observed in a Gaussian walk. This price is the one stored as input to the probability calculations.

Finally, price changes are computed using the Gaussian distribution. Random values according to this distribution are generated using the gaussianValue member function:

```
double OptionsProbabilities::gaussianValue(double mean, double sigma)
{
        std::normal_distribution<> distrib(mean, sigma);
        return distrib(random_generator);
}
```

To see how these functions are used together, consider the implementation of probFinishAboveStrike:

```
double OptionsProbabilities::probFinishAboveStrike()
{
    int nAbove = 0;
    for (int i=0; i<m_numIterations; ++i)
    {
        double val = getLastPriceOfWalk();
        if (val >= m_strike)
        {
            nAbove++;
        }
    }

    return nAbove/(double)m_numIterations;
}
```

The algorithm repeats as many iterations as are defined by the member variable m_numIterations. At each iteration, you request a new Gaussian random walk and store the last observed value. If the value satisfies the required property (in this case finishing above the strike price), then it is counted as an occurrence of the event. Finally, the member function returns the empirical probability defined by the percentage of favorable cases.

Complete Code

Listing 14-3 presents the random walk method to evaluate option probabilities. A sample main function is given at the end of the listing, showing how the OptionsProbabilities class can be invoked.

Listing 14-3. Class OptionsProbabilities

```
//
// OptionsProbabilities.h

#ifndef __FinancialSamples__OptionsProbabilities__
#define __FinancialSamples__OptionsProbabilities__

#include <vector>

class OptionsProbabilities {
public:
    OptionsProbabilities(double initialPrice, double strike, double
    avgStep, int nDays);
    OptionsProbabilities(const OptionsProbabilities &p);
    ~OptionsProbabilities();
    OptionsProbabilities &operator=(const OptionsProbabilities &p);

    void setNumIterations(int n);

    double probFinishAboveStrike();
    double probFinishBelowStrike();
    double probFinalPriceBetweenPrices(double lowPrice, double highPrice);
    std::vector<double> getWalk();
private:
    double m_initialPrice;
    double m_strike;
    double m_avgStep;
    int m_numDays;
    int m_numIterations;

    double gaussianValue(double mean, double sigma);
    double getLastPriceOfWalk();
};
```

```cpp
#endif /* defined(__FinancialSamples__OptionsProbabilities__) */
//
// OptionsProbabilities.cpp

#include "OptionsProbabilities.h"

#include <random>
+#include <iostream>

using std::vector;
using std::cout;
using std::endl;

static std::default_random_engine random_generator;

namespace  {
    const int NUM_ITERATIONS = 1000;
}

OptionsProbabilities::OptionsProbabilities(double initialPrice,
                                           double strike, double avgStep,
                                           int nDays)
: m_initialPrice(initialPrice),
  m_strike(strike),
  m_avgStep(avgStep),
  m_numDays(nDays),
  m_numIterations(NUM_ITERATIONS)
{
}

OptionsProbabilities::OptionsProbabilities(const OptionsProbabilities &p)
: m_initialPrice(p.m_initialPrice),
  m_strike(p.m_strike),
  m_avgStep(p.m_avgStep),
  m_numDays(p.m_numDays),
  m_numIterations(p.m_numIterations)
{
}
```

```cpp
OptionsProbabilities::~OptionsProbabilities()
{
}

OptionsProbabilities &OptionsProbabilities::operator=(const
OptionsProbabilities &p)
{
    if (this != &p)
    {
        m_initialPrice = p.m_initialPrice;
        m_strike = p.m_strike;
        m_avgStep = p.m_avgStep;
        m_numDays = p.m_numDays;
        m_numIterations = p.m_numIterations;
    }
    return *this;
}

void OptionsProbabilities::setNumIterations(int n)
{
    m_numIterations = n;
}

double OptionsProbabilities::probFinishAboveStrike()
{
    int nAbove = 0;
    for (int i=0; i<m_numIterations; ++i)
    {
        double val = getLastPriceOfWalk();
        if (val >= m_strike)
        {
            nAbove++;
        }
    }

    return nAbove/(double)m_numIterations;
}
```

```cpp
double OptionsProbabilities::probFinishBelowStrike()
{
    int nBelow = 0;
    for (int i=0; i<m_numIterations; ++i)
    {
        double val = getLastPriceOfWalk();
        if (val <= m_strike)
        {
            nBellow++;
        }
    }

    return nBelow/(double)m_numIterations;
}

double OptionsProbabilities::probFinalPriceBetweenPrices(double lowPrice,
double highPrice)
{
    int nBetween = 0;
    for (int i=0; i<m_numIterations; ++i)
    {
        double val = getLastPriceOfWalk();
        if (lowPrice <= val && val <= highPrice)
        {
            nBetween++;
        }
    }

    return nBetween/(double)m_numIterations;
}

double OptionsProbabilities::gaussianValue(double mean, double sigma)
{
        std::normal_distribution<> distrib(mean, sigma);
        return distrib(random_generator);
}

double OptionsProbabilities::getLastPriceOfWalk()
```

```cpp
{
    double prev = m_initialPrice;

    for (int i=0; i<m_numDays; ++i)
    {
        double stepSize = gaussianValue(0, m_avgStep);
        int r = rand() % 2;
        double val = prev;
        if (r == 0) val += (stepSize * val);
        else val -= (stepSize * val);
        prev = val;
    }
    return prev;
}

std::vector<double> OptionsProbabilities::getWalk()
{
    vector<double> walk;
    double prev = m_initialPrice;

    for (int i=0; i<m_numDays; ++i)
    {
        double stepSize = gaussianValue(0, m_avgStep);
        int r = rand() % 2;
        double val = prev;
        if (r == 0) val += (stepSize * val);
        else val -= (stepSize * val);
        walk.push_back(val);
        prev = val;
    }
    return walk;
}

int main()
{
    OptionsProbabilities optP(30, 35, 0.01, 100);

    cout << " above strike prob: "
```

```
        << optP.probFinishAboveStrike() << endl;
   cout << " below strike prob: "
        << optP.probFinishBelowStrike() << endl;
   cout << " between 28 and 32 prob: "
        << optP.probFinalPriceBetweenPrices(28, 32) << endl;

   return 0;
}
```

Running the Code

To run the code in Listing 14-3, you can use any standards-compliant C++ compiler such as gcc, llvm, or Visual C++. Once you compile the code and generate an executable file, the application can be run with the following results (exact numbers can vary depending on the random numbers used):

```
above strike prob: 0.055
below strike prob: 0.946
between 28 and 32 prob: 0.512
```

As you can see, the application is able to determine with good precision the probability that the price will finish above or below the strike. This is confirmed by the fact that the two first values add up to close to 100%. The approximation can still be improved by increasing the number of simulated random walks.

Conclusion

Monte Carlo methods are a general approach to problem solving that use randomization as a way to compute solutions that would be otherwise very difficult to find exactly. Due to the inherent randomness of financial markets, Monte Carlo methods appear as an important tool in the hands of the financial engineer and software developer.

You have seen in this chapter that such simulation techniques can be used to find quick solutions to diverse problems in the area of finance. For example, a common way to use Monte Carlo simulations is to forecast possible economic scenarios dictated by price variations. While this is a difficult task for traditional mathematical methods, one can easily design efficient algorithms such as a random walk. Such algorithms offer the ability to forecast prices using just a few input parameters based on past behavior.

In the first coding example, in Listing 14-1, we used a Monte Carlo technique to calculate the definite integral of a general function. While there are efficient ways to solve this problem with deterministic numeric algorithms, this problem shows the basic features of Monte Carlo methods and how their results can be interpreted and improved.

In Listing 14-2, you saw how to create a very simple random walk, which is one of the basic tools available for price simulation using Monte Carlo methods. You saw how to implement a version of random walk where price changes are uniformly distributed. You also saw a few common variations of the standard method, which are frequently used in applications.

Next, you learned about the use of Monte Carlo methods to calculate profit probabilities for options. The C++ class in Listing 14-3 illustrated a scheme that is easy to implement and can be used to analyze options and their possible profit scenarios. You have seen that using simulation and a few assumptions about price changes, one can easily determine the probability that a given stock will be in a certain price range within a number of days.

This chapter completes the discussion of the major mathematical tools used in financial software. In the next chapter, you will start to explore additional programming technologies that can be employed to support the creation and maintenance of such financial applications. You will see a number of examples that show how to integrate existing C++ code with other popular scripting languages, such as Python and Lua.

CHAPTER 15

Extending Financial Libraries

C++ is an expressive language that can be used to develop some of the most sophisticated software, including the high-performance applications that are routinely used in banks and other financial institutions. However, it is sometimes beneficial to combine and extend C++ libraries using interpreted languages that can simplify the creation of prototypes and other noncritical applications. A number of such interpreted languages are used for the purpose of connecting pre-compiled libraries. Among them, Python and Lua are among the most common interpreted languages employed in the financial industry.

In this chapter, I show how to use popular scripting and extension languages such as Lua and Python to interact with C++ libraries. The solutions and algorithms discussed in the next few sections allow you to reuse many of the same C++ components presented in previous chapters as part of applications developed in other languages. In some cases, you will also be able to use code that has been created in external languages in your own C++ applications.

The following are some of the topics discussed in this chapter:

- Extending C++ with Python: The Python language offers great features for the development of server-side applications. If you want to use C++ libraries as part of other services, Python might be the best way to integrate different libraries.

- Extending C++ with Lua: Lua is a relatively new language that is outstanding in its simple implementation of dynamic features. It is also used as an extension language that can be embedded into your own larger C++ applications.

© Carlos Oliveira 2021
C. Oliveira, *Practical C++20 Financial Programming*, https://doi.org/10.1007/978-1-4842-6834-6_15

Exporting C++ Stock Handling Code to Python

Generate code that provides the ability to export to Python a C++ stock handing class.

Solution

Python is a popular language that has been used in several domains, including web applications, scientific data exploration, and finance. One of the greatest strengths of Python is its ability to cleanly bring together large collections of programming libraries. A very important reason for that interoperability is the simple mechanism used by Python to interface with different languages, especially C and C++.

In this section, you will learn about the extension mechanism of Python and how it can be accessed from your C++ code. I have provided a financial example for this process, where you will give access to a Stock class originally designed and implemented in C++.

Python is available as open source on most operating systems. The main website where you can find the source code and binaries for most operating systems is http:// python.org. You will also find Python pre-installed on many UNIX-like operating systems, such as Linux and Mac OS X. The main mechanism for library extension in Python is the module. A module is used to export data and code to Python source files or other modules. The keyword import can be used to load an existing module into memory. For example:

```
import sys
```

This is a command used to load the sys module, which gives access to system-dependent functions. While modules can be created in Python itself, as C++ programmers our main interest is in creating modules using C++. This is possible with the Python module creating API (application programming interface), which is available for C and C++ and included with most Python installations.

Since the extension mechanism is written in C (for compatibility with existing C libraries for Python), it is necessary to create a number of C functions that encapsulate your original C++ classes in order to achieve interoperability with the Python environment.

As an example of how this process works, consider the Stock class, which can be used to model a single stock. Listing 15-1 shows the public interface for this class.

Listing 15-1. Interface for Stock Class

```
class Stock {
public:
    Stock(const std::string &ticker, double price, double div);
    Stock(const Stock &p);
    ~Stock();
    Stock &operator=(const Stock &p);

    std::string ticker();
    double price();
    void setPrice(double price);
    double dividend();
    void setDividend(double div);
    double dividendYield();
}
```

The extension API for Python is a set of header files and libraries that expose the Python environment to other applications written in C or C++. To export a class such as Stock to Python using the extension API, we need to create a few functions that will receive and return the requests sent by the Python interpreter. This is done in the Stock_Py.cpp file, which contains a list of functions that deal with each member of the Stock class. The first function of interest is the stock_create function, which is defined in the following way:

```
PyObject *stock_create(PyObject *self, PyObject *args)
```

It is common for functions called directly from Python to have a signature where two Python objects are received and a Python object is returned. The first argument to such a function is the Python object that is the target of the call (similar to the this pointer in C++) whenever the call is made using the syntax object.function(). The second parameter is a Python list that stores all the arguments passed to the function.

The first thing this function does is to retrieve the parameters passed as arguments. This can be done in the Python API by calling the function PyArg_ParseTuple, which is responsible for checking the arguments and copying their values into data objects. The first parameter of this function is the object representing the list of arguments. The second parameter is a string that defines the types of each data element in the argument list. Finally, the remaining arguments are pointers to the locations where the data should be stored.

```
if (!PyArg_ParseTuple(args, "sdd", &ticker, &price, &dividend))
        return NULL;
```

If this function is not successful, it will return false, which causes the `stock_create` to return `NULL`. Returning `NULL` indicates to the interpreter that the called function failed for some reason.

Once the input data has been validated, the next step is to create an object of class `Stock` and initialize it properly. The result is stored in a Python capsule object that is created with the function `PyCapsule_New`. This function call takes as the last parameter the name of a destructor function, which in this case is just `stock_destructor`, a function that calls the destructor for the `Stock` object. After the Python capsule has been created, the new Python object is then returned as the result of the function.

For an example of a function that just returns a single data object, consider `stock_ticker`, the function that gives access to the `Stock::ticker()` member function:

```
PyObject *stock_ticker(PyObject *self, PyObject *args)
{
    PyObject *obj;
    if (!PyArg_ParseTuple(args, "O!", &PyCapsule_Type, &obj))
        return NULL;

    Stock *stock = getStock(obj);
    return Py_BuildValue("s", stock->ticker().c_str());
}
```

In this function, the first step is to validate the input arguments, which uses the `PyArg_ParseTuple` function. The `PyArg_ParseTuple` function receives as arguments the container of the data and a string that determines the type of each argument, followed by pointers to variables where the data will be stored. The object to be retrieved in this case is of type `PyCapsule`, as defined in the remaining arguments (which give the object type and a pointer to an object variable). Once the argument is retrieved, you can use the `getStock` function to fetch the `Stock` pointer. Finally, the `ticker()` member function is called. To return the data to the Python interpreter, you need to convert the result into a Python object. This is done with the `Py_BuildValue` function, which uses a format string to determine the type of its remaining argument. Other functions are similar, and their main work is to retrieve data from the argument list and to convert the results into Python objects.

> **Note**   Even if a function that is exposed to Python returns no result, you're still required to return a valid Python object. In that case, you can use Py_None, which represents a standard Python object that means "no data."

Finally, we have the `initstock` function. It calls the `Py_InitModule` function from the Python API to determine the functions exposed in this module. The `Py_InitModule` function receives as parameters the name of the module and an array that contains a list of all functions (called stockMethods), their names, and descriptions. When this information is passed to the Python interpreter, it becomes available to Python developers whenever the `stock` module is imported.

Complete Code

Listing 15-2 shows the complete code for the `Stock` class and its associated Python glue code.

Listing 15-2. Class Stock Interface and Implementation

```
//
//  Stock.h

#ifndef __FinancialSamples__Stock__
#define __FinancialSamples__Stock__

#include <string>

class Stock {
public:
    Stock(const std::string &ticker, double price, double div);
    Stock(const Stock &p);
    ~Stock();
    Stock &operator=(const Stock &p);

    std::string ticker();
    double price();
    void setPrice(double price);
    double dividend();
```

```cpp
    void setDividend(double div);
    double dividendYield();
private:
    std::string m_ticker;
    double m_currentPrice;
    double m_dividend;
};

#endif /* defined(__FinancialSamples__Stock__) */


//
//  Stock.cpp

#include "Stock.h"

Stock::Stock(const std::string &ticker, double price, double div)
: m_ticker(ticker),
m_currentPrice(price),
m_dividend(div)
{
}

Stock::Stock(const Stock &p)
: m_ticker(p.m_ticker),
m_currentPrice(p.m_currentPrice),
m_dividend(p.m_dividend)
{
}

Stock::~Stock()
{
}

Stock &Stock::operator=(const Stock &p)
{
    if (this != &p)
    {
        m_ticker = p.m_ticker;
```

```cpp
        m_currentPrice = p.m_currentPrice;
        m_dividend = p.m_dividend;
    }
    return *this;
}

double Stock::price()
{
    return m_currentPrice;
}

void Stock::setPrice(double price)
{
    m_currentPrice = price;
}

double Stock::dividend()
{
    return m_dividend;
}

void Stock::setDividend(double div)
{
    m_dividend = div;
}

double Stock::dividendYield()
{
    return  m_dividend / m_currentPrice;
}

std::string Stock::ticker()
{
    return m_ticker;
}
//
//  Stock_Py.cpp
```

```cpp
#include "Stock_Py.h"
#include "Stock.h"

#include <Python.h>
#include <pycapsule.h>

#include <stdio.h>

namespace {

void stock_destructor(PyObject *capsule)
{
    printf("calling destructor\n");
    Stock *stock = reinterpret_cast<Stock*>(PyCapsule_GetPointer(capsule,
    NULL));
    delete stock;
}

PyObject *stock_create(PyObject *self, PyObject *args)
{
    char *ticker;
    double price;
    double dividend;
    if (!PyArg_ParseTuple(args, "sdd", &ticker, &price, &dividend))
        return NULL;

    printf("ticker: %s, price: %lf, dividend: %lf\n", ticker, price,
    dividend);

    Stock *stock = new Stock(ticker, price, dividend);

    PyObject* stockObj = PyCapsule_New(stock, NULL,  stock_destructor);
    return stockObj;
}

Stock *getStock(PyObject *obj)
{
    if (!PyCapsule_CheckExact(obj))
        printf("error: not a stock object\n");
```

```
    fflush(stdout);
    return reinterpret_cast<Stock*>(PyCapsule_GetPointer(obj, NULL));
}

PyObject *returnNone()
{
    Py_INCREF(Py_None);
    return Py_None;
}

PyObject *stock_ticker(PyObject *self, PyObject *args)
{
    PyObject *obj;
    if (!PyArg_ParseTuple(args, "O!", &PyCapsule_Type, &obj))
        return NULL;

    Stock *stock = getStock(obj);
    return Py_BuildValue("s", stock->ticker().c_str());
}

PyObject *stock_price(PyObject *self, PyObject *args)
{
    PyObject *obj;
    if (!PyArg_ParseTuple(args, "O!", &PyCapsule_Type, &obj))
        return NULL;

    Stock *stock = getStock(obj);
    return Py_BuildValue("d", stock->price());
}

PyObject *stock_setPrice(PyObject *self, PyObject *args)
{
    double price;
    PyObject *obj;
    if (!PyArg_ParseTuple(args, "O!d", &PyCapsule_Type, &obj, &price))
        return NULL;

    Stock *stock = getStock(obj);
    if (!stock)
```

```
        return NULL;
    stock->setPrice(price);
    return returnNone();
}

PyObject *stock_dividend(PyObject *self, PyObject *args)
{
    PyObject *obj;
    if (!PyArg_ParseTuple(args, "O!", &PyCapsule_Type, &obj))
        return NULL;

    Stock *stock = getStock(obj);
    if (!stock)
        return NULL;
    return Py_BuildValue("d", stock->dividend());
}

PyObject *stock_setDividend(PyObject *self, PyObject *args)
{
    double dividend;
    PyObject *obj;
    if (!PyArg_ParseTuple(args, "O!d", &PyCapsule_Type, &obj, &dividend))
        return NULL;

    Stock *stock = getStock(obj);

    stock->setDividend(dividend);
    return returnNone();
}

PyObject *stock_dividendYield(PyObject *self, PyObject *args)
{
    PyObject *obj;
    if (!PyArg_ParseTuple(args, "O!", &PyCapsule_Type, &obj))
        return NULL;
    Stock *stock = getStock(obj);
    return Py_BuildValue("d", stock->dividendYield());
}
```

```
PyMethodDef stockMethods[] = {
    {"new",  stock_create, METH_VARARGS, "Create a new stock object."},
    {"ticker",  stock_ticker, METH_VARARGS, "get ticker for a stock object."},
    {"price",  stock_price, METH_VARARGS, "get price for stock."},
    {"setPrice",  stock_setPrice, METH_VARARGS, "set price for a stock object."},
    {"dividend",  stock_dividend, METH_VARARGS, "get dividend for stock."},
    {"setDividend",  stock_setDividend, METH_VARARGS, "set dividend for a
    stock object."},
    {"dividendYield",  stock_dividendYield, METH_VARARGS, "get dividend
    yield for stock."},
    {NULL, NULL, 0, NULL}
};

}

PyMODINIT_FUNC initstock()
{
    Py_InitModule("stock", stockMethods);
}
#
# stock-setup.py

from distutils.core import setup, Extension

setup(name="stock", version="1.0",
      ext_modules=[Extension("stock", ["Stock.cpp", "Stock_Py.cpp"])])
```

Running the Code

The process of building a Python extension module is a little different from what
you did for other C++ applications described in this book. The process requires the
creation of a loadable module, which has a different form in each platform, such as a
dll (on Windows) or shared object file on many UNIX systems. To simplify this process,
the Python developers created a tool that uses a `setup.py` file to perform the build

automatically. Using setup.py, you don't need to figure out particular build information such as including directories and linking libraries. As you can see in Listing 15-1, the file stock-setup.py describes the extension, with the source files necessary to build the module. The building process (executed on a Mac OS X system) is shown as follows. As seen in the listing, the C++ building is generated automatically by Python using the build_ext option.

```
$ python stock-setup.py build_ext -i
running build_ext
building 'stock' extension
cc -fno-strict-aliasing -fno-common -dynamic -arch x86_64 -arch i386 -g -Os
-pipe
-fno-common -fno-strict-aliasing -fwrapv -DENABLE_DTRACE -DMACOSX -DNDEBUG
-Wall
-Wstrict-prototypes -Wshorten-64-to-32 -DNDEBUG -g -fwrapv -Os -Wall
-Wstrict-prototypes -DENABLE_DTRACE -arch x86_64 -arch i386 -pipe -I/
System/Library/Frameworks/Python.framework/Versions/2.7/include/python2.7
-c Stock.cpp -o build/temp.macosx-10.9-intel-2.7/Stock.o
cc -fno-strict-aliasing -fno-common -dynamic -arch x86_64 -arch i386 -g -Os
-pipe
-fno-common -fno-strict-aliasing -fwrapv -DENABLE_DTRACE -DMACOSX -DNDEBUG
-Wall
-Wstrict-prototypes -Wshorten-64-to-32 -DNDEBUG -g -fwrapv -Os -Wall
-Wstrict-prototypes -DENABLE_DTRACE -arch x86_64 -arch i386 -pipe -I/
System/Library/Frameworks/Python.framework/Versions/2.7/include/python2.7
-c Stock_Py.cpp -o build/temp.macosx-10.9-intel-2.7/Stock_Py.o
c++ -bundle -undefined dynamic_lookup -arch x86_64 -arch i386 -Wl,-F.
build/temp.macosx-10.9-intel-2.7/Stock.o build/temp.macosx-10.9-intel-2.7/
Stock_Py.o -o /Users/carlosoliveira/code/FinancialSamples/FinancialSamples/
stock.so
```

Once the module is compiled, it can be easily loaded into a Python script or iterative session. The following is a transcript of a sample use of the stock module:

```
$ python
Python 2.7.5 (default, Mar 9 2014, 22:15:05)
[GCC 4.2.1 Compatible Apple LLVM 5.0 (clang-500.0.68)] on darwin
```

```
Type "help", "copyright", "credits" or "license" for more information.
>>> import stock
>>> a = stock.new('IBM',1,1)
ticker: IBM, price: 1.000000, dividend: 1.000000
>>> stock.setPrice(a, 105)
>>> stock.price(a)
105.0
>>> stock.setDividend(a, 2.2)
>>> stock.dividend(a)
2.2
>>> stock.dividendYield(a)
0.020952380952380955
```

Exporting C++ Classes Directly to Python

Write C++ code to export existing classes into Python applications.

Solution

In the section "Exporting C++Stock Handling Code to Python," you saw how C++ code can be exported to Python using the module mechanism. Through the external Python API, it is possible to expose functions and classes that were previously created using C++. However, the external API also imposes the creation of glue code that is not only a boring task but also an error-prone job, which could be better done by a computer.

To simplify some of the issues raised by the external Python API, a new boost library was developed. The boost::python library uses a template-based mechanism to automatically create the integration code required by Python. In this way, developers can more easily expose classes, variables, and function using a set of C++ templates while avoiding repetitive tasks such as converting data from and to Python objects. The following are a few advantages of using boost to export C++ code to Python:

- Avoid boilerplate: A lot of the code necessary to export C++ classes into Python modules is simple and repetitive. Using a template solution makes it easy to reduce or totally remove much of the boilerplate code needed by the Python external API.

413

- Provide type safety: In Python, objects have all the same type PyObject. While there is runtime checking for the correct type in Python, you should be able to use C++ compile-time checking whenever possible. With boost::python, C++ types are used, and conversion into Python is done automatically and only when needed.

- Reduce programming effort: Using the boost library, you can leverage a lot of code that has been developed to solve the specific problem of exporting C++ classes to Python. By using the Python API directly, you may encounter problems that have already been solved in boost::python. As in other areas of C++ programming, the ideal is to reuse good libraries and designs instead of reinventing existing solutions.

Despite the advantages of boost::python, there are also some reasons why you may want to avoid it and use the Python API directly:

- Size of the project: Sometimes, you need it to export only a single class or function to Python for a special use. In that case, it may be just as easy to stick to the Python API and skip boost::python.

- Boost integration: If your project doesn't use boost, or if you're not allowed to incorporate other boost libraries into your code, then it would be difficult to use this solution for Python integration.

- Special needs of the project: While the templates in boost::python are very flexible, you may have some additional requirements for the types you're exporting. In that case, only the underlying Python extension API may provide the flexibility needed by your project.

Using the boost::python library is straightforward, and you just need to look at some examples and its reference to understand how to quickly export classes. In the code presented in Listing 15-3, I will show how to do this for a single example, the Matrix class from Chapter 5.

The boost::python library is installed along with other boost libraries, so if boost is already installed in your system, you should be ready to use it. The main header file for the library is <boost/python.hpp>, which gives access to all macros and templates necessary to export C++ classes.

The main macro in the library is BOOST_PYTHON_MODULE. This macro is needed to generate the boilerplate code that the Python runtime expects. In the scope that follows the macro, you can declare the classes, functions, and other types that will be visible from Python.

To export C++ classes, the main facility provided is the class_ template. As you would expect, this template is used to perform the hard work of defining types and their properties using the underlying Python API. The template parameter for class_ is the class name. The constructor requires a name to be used by Python and a default constructor. Constructors are defined using the init template, with the parameters added as template parameters.

Attached to the main class_ template, you will find calls to the def member function, which is used to define new members to the class. Every time you call def, the class_ template generates additional code to handle calls from Python code into a given member function. So, for example, you have the following definition:

```
def("subtract", &MatrixP::subtract)
```

Here, the subtract member function is defined with the name listed as the first argument and the destination of the call listed as the second argument.

Complete Code

You can see the complete code for the Matrix module in Listing 15-3. The setup.py file at the end of the listing can be used to build the module.

Listing 15-3. Matrix Module and Associated setup.py File

```
//
// Matrix_Py.h

#ifndef __FinancialSamples__Matrix_Py__
#define __FinancialSamples__Matrix_Py__

#include <iostream>

#include "Matrix.h"

class MatrixP : public Matrix {
public:
```

```
    MatrixP(int a);
    MatrixP(int a, int b);
    MatrixP(const MatrixP &p);
    ~MatrixP();
    void set(int a, int b, double v);
    double get(int a, int b);
};

#endif /* defined(__FinancialSamples__Matrix_Py__) */

//
//  Matrix_Py.cpp

#include "Matrix_Py.h"

// include this header file for access to boost::python templates and macros
#include <boost/python.hpp>

// add the using clause to reduce namespace clutter
using namespace boost::python;

MatrixP::MatrixP(int a)
: Matrix(a)
{

}

MatrixP::MatrixP(int a, int b)
: Matrix(a, b)
{

}

MatrixP::MatrixP(const MatrixP &p)
: Matrix(p)
{

}
```

```
MatrixP::~MatrixP()
{
}

void MatrixP::set(int a, int b, double v)
{
    (*this)[a][b] = v;
}

double MatrixP::get(int a, int b)
{
    return (*this)[a][b];
}

// this macro generates all the boilerplate required by the Python API
BOOST_PYTHON_MODULE(matrix)
{

    // defines a new class to be exported
    class_<MatrixP>("Matrix",
                    init<int>())    // the init form defines a constructor

        // another constructor with two int parameters
        .def(init<int, int>())

        // here are some standard functions (name first, member function
        second)
        .def("add", &MatrixP::add)
        .def("subtract", &MatrixP::subtract)
        .def("multiply", &MatrixP::multiply)
        .def("numRows", &MatrixP::numRows)
        .def("trace", &MatrixP::trace)
        .def("transpose", &MatrixP::transpose)
        .def("set", &MatrixP::set)
        .def("get", &MatrixP::get)
    ;
}
#
```

```
# matrix-setup.py
#
# python code to build the matrix module
from distutils.core import setup, Extension

# you need to include include and library paths for the boost::python
library
setup(name="matrix", version="1.0",
      ext_modules=[Extension("matrix", ["Matrix.cpp", "matrix_Py.cpp"],
                             include_dirs=["/opt/local/include/"],
                             library_dirs=["/opt/local/lib/"],
                             libraries=["boost_python-mt"])])
```

Running the Code

To compile the code, you can follow a procedure similar to the one described in the previous section. This means that you can use the Python build system to compile the extension (usually into a .so or .dll format). To do this, you need to create a setup file, which in our case is listed as matrix-setup.py. Notice that the libraries key is also listed in the matrix-setup.py file. This key tells the build system to link against the boost_python-mt library. You may also need to change the include_dirs and the library_dirs keys to the location where boost is installed in your system. The following is the result of running the setup file through Python:

```
$ python matrix-setup.py build_ext -i
running build_ext
building 'matrix' extension
cc -fno-strict-aliasing -fno-common -dynamic -arch x86_64 -arch i386 -g -Os
-pipe
-fno-common -fno-strict-aliasing -fwrapv -DENABLE_DTRACE -DMACOSX -DNDEBUG
-Wall
-Wstrict-prototypes -Wshorten-64-to-32 -DNDEBUG -g -fwrapv -Os -Wall
-Wstrict-prototypes -DENABLE_DTRACE -arch x86_64 -arch i386 -pipe -I/opt/
local/include/ -I/System/Library/Frameworks/Python.framework/Versions/2.7/
include/python2.7 -c Matrix.cpp -o build/temp.macosx-10.9-intel-2.7/
Matrix.o
```

```
c++ -bundle -undefined dynamic_lookup -arch x86_64 -arch i386 -Wl,-F.
build/temp.macosx-10.9-intel-2.7/Matrix.o build/temp.macosx-10.9-intel-2.7/
matrix_Py.o -L/opt/local/lib/ -lboost_python-mt -o /Users/carlosoliveira/
code/FinancialSamples/FinancialSamples/matrix.so
```

After this process is finished, you should have a file called matrix.so (or matrix.dll, if you're building on a Windows system). You can load it with an import statement like the following:

```
$ python
Python 2.7.5 (default, Mar 9 2014, 22:15:05)
[GCC 4.2.1 Compatible Apple LLVM 5.0 (clang-500.0.68)] on darwin
Type "help", "copyright", "credits" or "license" for more information.
>>> import matrix
>>> m = matrix.Matrix(5,5)
>>> m.set(2,2,4)
>>> m.get(2,2)
4.0
```

Using Lua as an Extension Language

Use Lua as an extension library for classes written in C++.

Solution

Lua is a scripting language that was designed to provide extension mechanisms for C and C++ code and to work as an embedded language for other applications. In this respect, it has been very successful, with a large number of software products that currently use Lua to implement extension modules based on existing C and C++ class libraries. Examples of such uses can be found in computer games, image-processing packages, and software for the financial industry.

The success of Lua is linked to its simple system, which tries to mix as closely as possible with the C and C++ environment. With this goal in mind, Lua offers only the basic mechanisms necessary to build a dynamic, garbage-collected runtime system. These features of the language have made it an easy choice for the creation of programmatic extensions to large-scale application code bases.

You can download Lua in its source form from the main website http://lua.org. The easiest way to integrate Lua into a C++ project is to add the source files directly. You can also decide to create a separate library containing the Lua interpreter and link it to your application, using information made available in the Lua documentation. To use Lua as an extension language, the first step is to import and initialize the Lua runtime engine. This can be done using the Lua C API, which is part of the standard installation of the language. The main header file, lua.h, gives developers access to the main features of the runtime engine, as well as to Lua's standard library.

The following is the main function for the example application, where you can see the sequence of operations necessary to load Lua into your program:

```
int main (void) {
    char buff[256];

    lua_State *L = luaL_newstate();
    int error;

    // load some of the (C) libraries included with Lua
    luaopen_base(L);
    luaopen_table(L);
    luaopen_io(L);
    luaopen_string(L);
    luaopen_math(L);

    // load LuaOption object
    LuaWrapper<LuaOption>::Register(L);

    while (fgets(buff, sizeof(buff), stdin) != NULL) {
        error = luaL_loadbuffer(L, buff, strlen(buff), "line") ||
        lua_pcall(L, 0, 0, 0);
        if (error) {
            cerr << lua_tostring(L, -1) << endl;
            lua_pop(L, 1);  // remove error from Lua stack
        }
    }

    lua_close(L);
    return 0;
}
```

Most of the functions in this API use the lua_State structure as a parameter. As you can see in the aforementioned example, you can create a new lua_State object using the luaL_newstate function. Once this initialization step has been completed, you can load some of the libraries included with the Lua runtime. It is possible to choose subsets of the library using functions such as luaopen_string and luaopen_math, which load Lua functions to handle strings and math operations, respectively.

The next step is to load any user-defined libraries that you might need into Lua data tables. The Lua language is organized in such a way that its dynamic information is stored in a stack, that is, a first-in/first-out data structure. There is a global stack, where the equivalent of global variables is stored. You can use this stack to store new tables as necessary. To store individual values, you push them into the stack using functions such as lua_pushstring, lua_pushnumber, or lua_pushclosure (for functions). You can read data from the top of the stack using functions such as lua_tonumber.

Another thing that can be done with the Lua runtime is to directly call one of the Lua functions. To call a function, you need to push the name of the function you want to call into the stack, followed by the required parameters. Next, you need to call the function lua_pcall, which performs the call. You can see an example for lua_pcall at the end of the main function presented previously. Finally, you can access the results of the function, which on return are stored at the top of the stack.

While this initially seems to be a lot of work, it can be done easily because of the generic nature of the Lua extension API. I provide an example of how to access a C++ class from Lua code using the Option class, which contains just two data members: a ticker (string) and the strike price (double). The original class is accessed from Lua using the class LuaOption. The reason this is necessary has to do with the fact that Lua can interact only with functions that receive a parameter of type lua_State. Each method of LuaOption retrieves the data from the stack, calls the corresponding method in the Option class, and returns the results in the stack. Finally, the LuaOption class is registered with the help of the template LuaWrap.

Complete Code

The example in Listing 15-4 shows how to use the Lua API to embed an extension language into your application. The only class exposed in this example is the Option class. The class LuaOption is a simple wrapper for Option, and it is responsible for converting parameters from and to Lua types. The main function has the ability of loading a Lua file and calling any functions contained in it.

Listing 15-4. Class Option and Its Lua Wrapper LuaOption

```
//
//  Option.h

#ifndef __FinancialSamples__Option__
#define __FinancialSamples__Option__

#include <string>

class Option {
public:
    Option(const std::string &ticker, double strike);
    Option(const Option &p);
    ~Option();
    Option &operator=(const Option &p);

    std::string ticker();
    double strike();

    void setTicker(const std::string &);
    void setStrike(double);

private:
    std::string m_ticker;
    double m_strike;

};

#endif /* defined(__FinancialSamples__Option__) */

//
//  Option.cpp

#include "Option.h"

Option::Option(const std::string &ticker, double strike)
: m_ticker(ticker),
  m_strike(strike)
{
}
```

```
Option::Option(const Option &p)
: m_ticker(p.m_ticker),
  m_strike(p.m_strike)
{
}

Option::~Option()
{
}
.
Option &Option::operator=(const Option &p)
{
    if (this != &p)
    {
        m_ticker = p.m_ticker;
        m_strike = p.m_strike;
    }
    return *this;
}

std::string Option::ticker()
{
    return m_ticker;
}

double Option::strike()
{
    return m_strike;
}

void Option::setTicker(const std::string &s)
{
    m_ticker = s;
}

void Option::setStrike(double val)
{
    m_strike = val;
}
```

```cpp
//
//  LuaOption.h

#ifndef __FinancialSamples__LuaOption__
#define __FinancialSamples__LuaOption__

#include "LuaWrap.h"

class Option;

#include <string>

class LuaOption {
public:
    LuaOption(lua_State *l);
    void setObject(lua_State *l);

    static const char className[];
    static LuaWrapper<LuaOption>::RegType methods[];

    // Lua functions should receive lua_State and return int
    int ticker(lua_State *l);
    int strike(lua_State *l);

    int setTicker(lua_State *l);
    int setStrike(lua_State *l);
private:
    Option *m_option;
};

#endif /* defined(__FinancialSamples__LuaOption__) */

//
//  LuaOption.cpp

#include "LuaOption.h"
#include "Option.h"

#include <lauxlib.h>

const char LuaOption::className[] = "Option";
```

```cpp
LuaOption::LuaOption(lua_State *L)
{
    m_option = (Option*)lua_touserdata(L, 1);
}

void LuaOption::setObject(lua_State *L)
{
    m_option = (Option*)lua_touserdata(L, 1);
}

int LuaOption::ticker(lua_State *L)
{
    lua_pushstring(L, m_option->ticker().c_str());
    return 1;
}

int LuaOption::strike(lua_State *L)
{
    lua_pushnumber(L, m_option->strike());
    return 1;
}

int LuaOption::setTicker(lua_State *L)
{
    m_option->setTicker((const char*)luaL_checkstring(L, 1));
    return 0;
}

int LuaOption::setStrike(lua_State *L)
{
    m_option->setStrike((double)luaL_checknumber(L, 1));
    return 0;
}

#define method(class, name) {#name, &class::name}
LuaWrapper<LuaOption>::RegType LuaOption::methods[] = {
    method(LuaOption, ticker),
    method(LuaOption, strike),
```

```
    method(LuaOption, setTicker),
    method(LuaOption, setStrike),
    {0,0}
};

//
//  LuaWrapper.h
// original code from luna wrapper example (from http://lua-users.org/wiki/
LunaWrapper)

#ifndef __FinancialSamples__Luna__
#define __FinancialSamples__Luna__

#include <lua.h>

template<class T> class LuaWrapper {
public:
    static void Register(lua_State *L) {
        lua_pushcfunction(L, &LuaWrapper<T>::constructor);
        lua_setglobal(L, T::className);

        luaL_newmetatable(L, T::className);
        lua_pushstring(L, "__gc");
        lua_pushcfunction(L, &LuaWrapper<T>::gc_obj);
        lua_settable(L, -3);
    }

    static int constructor(lua_State *L) {
        T* obj = new T(L);

        lua_newtable(L);
        lua_pushnumber(L, 0);
        T** a = (T**)lua_newuserdata(L, sizeof(T*));
        *a = obj;
        luaL_getmetatable(L, T::className);
        lua_setmetatable(L, -2);
        lua_settable(L, -3); // table[0] = obj;

        for (int i = 0; T::methods[i].name; i++) {
```

```
            lua_pushstring(L, T::methods[i].name);
            lua_pushnumber(L, i);
            lua_pushcclosure(L, &LuaWrapper<T>::thunk, 1);
            lua_settable(L, -3);
        }
        return 1;
    }.
    static int thunk(lua_State *L) {
        int i = (int)lua_tonumber(L, lua_upvalueindex(1));
        lua_pushnumber(L, 0);
        lua_gettable(L, 1);

        T** obj = static_cast<T**>(luaL_checkudata(L, -1, T::className));
        lua_remove(L, -1);
        return ((*obj)->*(T::methods[i].mfunc))(L);
    }

    static int gc_obj(lua_State *L) {
        T** obj = static_cast<T**>(luaL_checkudata(L, -1, T::className));
        delete (*obj);
        return 0;
    }

    struct RegType {
        const char *name;
        int(T::*mfunc)(lua_State*);
    };
};

#endif /* defined(__FinancialSamples__LuaWrapper__) */

//
//  LuaMain.cpp

#include "LuaMain.h"

#include <iostream>
```
.

```cpp
#include <string.h>
#include <lua.h>
#include <lauxlib.h>
#include <lualib.h>

using std::cout;
using std::cerr;
using std::endl;

int main (void) {
    char buff[256];

    lua_State *L = luaL_newstate();
    int error;

    // load some of the (C) libraries included with Lua
    luaopen_base(L);
    luaopen_table(L);
    luaopen_io(L);
    luaopen_string(L);
    luaopen_math(L);

    // load LuaOption object
    LuaWrapper<LuaOption>::Register(L);

    while (fgets(buff, sizeof(buff), stdin) != NULL) {
        error = luaL_loadbuffer(L, buff, strlen(buff), "line") ||
        lua_pcall(L, 0, 0, 0);
        if (error) {
            cerr << lua_tostring(L, -1) << endl;
            lua_pop(L, 1);  // remove error from Lua stack
        }
    }

    lua_close(L);
    return 0;
}.
```

Running the Code

.You can build the code in Listing 15-4 using any standards-compliant C++ compiler. The only additional step is that you will have to add the path for the Lua including files and libraries in the compiler configuration. For example, in my system, I compiled the Lua files in the directory /code/lua-5.2.3/src/, so this code can be built as follows:

```
$ c++ -o luatest Option.cpp LuaOption.cpp  LuaMain.cpp -I/Users/code/lua-
5.2.3/src \
    -L/Users /code/lua-5.2.3/src/ -llua
```

Conclusion

Extension languages such as Python and Lua have become very popular in the last few years. They provide the ability to quickly develop applications that are composed of existing components. Thanks to C++ flexibility, however, as you have seen, it is possible to create C++ libraries that can be easily integrated with these languages.

Initially, you have seen how to use Python as an extension language for C++ classes. The class presented as an example can be accessed directly from Python code by simply using the Python external API. You have seen how to convert data from and into the data structures maintained by Python. In a second coding example, you learned how to use the boost::python library, which provides a more concise way to export C++ data types to Python. I have discussed some of the advantages and disadvantages of each method.

Lua is another language that has grown in popularity in the last few years. With its small footprint, Lua is an ideal candidate for the position of extension language for libraries written in C++. Due to its simplicity and modularity, you can easily embed the Lua interpreter in a C++ application. In this chapter, you saw a C++ coding technique that shows how to easily integrate Lua into your applications.

Using your C++ code as an external library is one of the many ways you can connect with other tools and environments. Another option is to integrate your financial C++ code into existing scientific programming tools. Two of the most often used scientific tools for data analysis are R and Maxima. In Chapter 16, you will learn more about these tools and how to integrate C++ into the workflow provided by these applications.

CHAPTER 16

Using C++ with R and Maxima

One of the advantages of code implemented in C++ is that it can be used as part of native libraries or stand-alone applications and also integrated as a component of other development and modeling environments. In the financial industry, for instance, it is common to have lower-level modules implemented in C++, with high-level analysis being performed in more user-oriented environments such as Excel, Mathematica, Matlab, Maxima, R, and Octave.

When using a high-level data analysis environment, it is crucial to have numeric results that are identical to the ones achieved in native code and to access the same underlying libraries that are already coded in C++. For this reason, an important skill for programmers working in the financial industry is to be able to integrate existing code with one or more of the analytical applications used by analysts and mathematicians.

In this chapter, we show you how to incorporate financial libraries developed in C++ into two well-known simulation and modeling environments for financial analysis: R and Maxima. These are open source applications that are freely available on multiple platforms. However, the examples you see in this chapter demonstrate principles that can be applied to other commercial tools in the areas of statistics, simulation, engineering, and mathematics.

The following are a few topics that you will learn about in this chapter:

- Integrating C++ with R: Users of the R language have created a rich ecosystem of statistical libraries and applications. However, it is sometimes necessary to integrate C++ code as part of the analysis performed in R. You will see in this chapter how to easily embed C++ classes into this system, both for increased performance as well as for consistency with other applications deployed in C++.

© Carlos Oliveira 2021
C. Oliveira, *Practical C++20 Financial Programming*, https://doi.org/10.1007/978-1-4842-6834-6_16

- Integrating C++ with Maxima: The Maxima Computer Algebra System is used to develop precise mathematical models with a simple, high-level language. It is also used for its visualization facilities. You can easily integrate existing C++ libraries into Maxima using the shared library mechanism supported by the language.

Integrating C++ with R

Create a C++ class to calculate the present value of a set of payments and that can be called from the R interpreter.

Solution

R is a programming environment that was created to perform statistical analysis of large data sets. Due to its easy-to-use and advanced statistical abilities, R has become the most used environment for data analysis and is the de facto standard in some areas such as data mining. A growing number of statisticians and engineers use R daily to study the properties of large data sets.

R is available for the most common operating systems and computer architectures. You can download it for free from the official website at www.r-project.org. After running the required installation method for your operating system, you will be able to start the iterative interpreter for the language. The standard R environment is able to run R scripts and single commands. You can use these tools to perform quick data analysis and create plots based on existing data. You can see in Figure 16-1 what the main application window for the R console looks like.

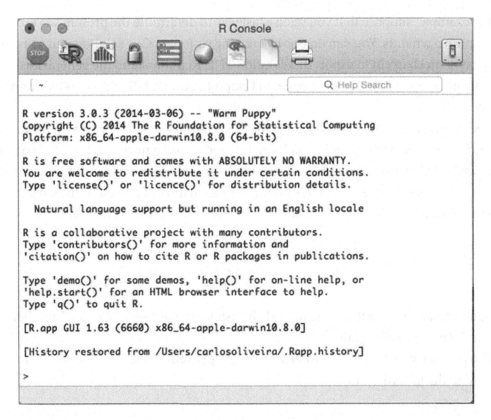

Figure 16-1. *R console window on Mac OS X*

R is typically used to perform a large range of mathematical algorithms and data analysis tasks. For example, a common use of R is to run standard statistical procedures, such as mean squared error and other types of statistical regression. R has also been used to implement statistical tests tailored for financial data sets. As a consequence, it can be very useful to be able to load C++ code into the dynamic environment provided in the R console.

To make it possible to use C++ classes in R, you need to employ the R extension application programming interface (API). The extension API consists of a set of C-based functions that interact with the R runtime. For example, you can use the API to retrieve and convert values from R. Similarly, you can use the API for common tasks such as calling mathematical functions and random number generators, among others.

To create C++ functions and classes that access the R extension library, you need to include the header file R.h. This header is the main C file that coordinates access to the many API declarations exported by the R runtime.

For example, suppose that you need a fast way to determine the present value of a set of future payments. You can do this by creating your own C++ solution and using the R extension mechanism to export this solution to R. To understand the general idea of present-value calculations, you can refer to Chapter 1, where I discussed tools for fixed income analysis.

The main functionality is encoded in the REXtension class. The two main methods in this class are addCashPayment, which can be used to add a new cash flow that will be later considered by the algorithm, and presentValue, which calculates the present value of all cash payments added up to this point. The calculation of present value uses the following formula:

$$\sum_{i=1}^{N} \frac{V_i}{(1+R)^{T_i}}$$

In this equation, V_i is the value of the i-th cash flow, T_i is the time of the i-th cash flow, and R is the interest rate.

The real entry point for the R interpreter is the presentValue function, which is declared in the following way:

```
extern "C" {
void presentValue(int *, double *, int *, double *, double *);
}
```

The reason for the extern "C" statement in this declaration is to avoid the normal mangling of function names that is performed by the C++ compiler. Only the name of the function is affected, and the contents of the function can use most C++ features. By declaring the function in such a way, the presentValue name will be maintained in the library without modifications, so that the R interpreter can view and access it.

The definition of the presentValue function is not unusual. The only difference to normal C++ code is that all parameters are passed as pointers. This is the way in which the R runtime allows data to be shared between the interpreter and the C++ code. Using pointers, the called function can both read and modify the passed arguments, if necessary. The implementation of the function uses the information passed in the parameters, which include the number of elements in the cash flow vector, the interest rate, then a vector with time indication, followed by a vector of cash flows. The last parameter is a pointer to the result value.

Complete Code

Listing 16-1 shows the implementation of class REextension. The main parts are the definition of the class, which contains the functionality to calculate the present value of a set of cash flows, and the presentValue function, which can be directly accessed from the R runtime.

Listing 16-1. Code for the R Extension Library

```
//
// RExtension.h

#ifndef __FinancialSamples__RExtension__
#define __FinancialSamples__RExtension__

#include <vector>

class RExtension {
public:
    RExtension(double rate);
    RExtension(const RExtension &p);
    ~RExtension();
    RExtension &operator=(const RExtension &p);

    void addCashPayment(double value, int timePeriod);
    double presentValue();
private:
    std::vector<double> m_cashPayments;
    std::vector<int> m_timePeriods;
    double m_rate;
    double presentValue(double futureValue, int timePeriod);
};

#endif /* defined(__FinancialSamples__RExtension__) */

//
// RExtension.cpp

#include "RExtension.h"
```

```cpp
#include <iostream>
#include <cmath>

using std::cout;
using std::endl;

extern "C" {
void presentValue(int *, double *, int *, double *, double *);
}

void presentValue(int *numPayments, double *intRate,
                  int *timePeriods, double *payments, double *result)
{
    int n = *numPayments;
    RExtension re(*intRate);
    for (int i=0; i<n; ++i)
    {
        re.addCashPayment(payments[i], timePeriods[i]);
    }
    *result = re.presentValue();
}

RExtension::RExtension(double rate)
: m_rate(rate)
{
}

RExtension::RExtension(const RExtension &v)
: m_rate(v.m_rate)
{
}

RExtension::~RExtension()
{
}

RExtension &RExtension::operator =(const RExtension &v)
```

```cpp
{
    if (this != &v)
    {
        this->m_cashPayments = v.m_cashPayments;
        this->m_timePeriods = v.m_timePeriods;
        this->m_rate = v.m_rate;
    }
    return *this;
}

void RExtension::addCashPayment(double value, int timePeriod)
{
    m_cashPayments.push_back(value);
    m_timePeriods.push_back(timePeriod);
}

double RExtension::presentValue(double futureValue, int timePeriod)
{
    double pValue = futureValue / pow(1+m_rate, timePeriod);
    cout << " value " << pValue << endl;
    return pValue;
}

double RExtension::presentValue()
{
    double total = 0;
    for (unsigned i=0; i<m_cashPayments.size(); ++i)
    {
        total += presentValue(m_cashPayments[i], m_timePeriods[i]);
    }
    return total;
}
```

Running the Code

The code presented in the previous section can be built using any standards-compliant C++ compiler. However, the R interpreter makes it very easy to create an extension library using the CMD option of the R command. This build technique allows you to quickly create a shared object file that contains all the necessary C++ code, in a way that can be readily imported into the R runtime. Using the CMD option of the R interpreter, you also don't need to worry about the correct compiler, the right location of header files, and link libraries, as well as other common compilation parameters.

Here is how I generated a binary object from the given source file (this was run on a Mac OS X system, but the result should be similar in other platforms). The compilation command is automatically generated by R using the SHLIB option, so you don't need to worry about the locations of libraries.

```
$ R CMD SHLIB code/FinancialSamples/FinancialSamples/RExtension.cpp
g++ -arch x86_64 -I/Library/Frameworks/R.framework/Resources/include
-DNDEBUG
-I/usr/local/include     -fPIC  -mtune=core2 -g -O2  -c
code/FinancialSamples/FinancialSamples/RExtension.cpp -o
code/FinancialSamples/FinancialSamples/RExtension.o
g++ -arch x86_64 -dynamiclib -Wl,-headerpad_max_install_names -undefined
dynamic_lookup -single_module -multiply_defined suppress -L/usr/local/lib
-L/usr/local/lib -o
code/FinancialSamples/FinancialSamples/RExtension.so
code/FinancialSamples/FinancialSamples/RExtension.o -F/Library/
Frameworks/R.framework/.. -framework R -Wl,-framework -Wl,CoreFoundation
```

Once the shared object has been created (this is a file with a .so extension on UNIX or .dll extension on Windows), you can load it into the R interpreter using the dyn. load function with the name of the file as the single parameter. After that, you just need to use the .C function to call the compiled C or C++ function. For this to work, you need to provide the function name as the first argument, followed by the arguments to the function. You need to ensure, however, that the values passed to the function are marked with the right parameter types (using functions such as as.integer and as.double). After the .C function is executed, the resulting values are printed in the interpreter window. Here is a sample session, where I import and use the RExtension module.

```
$ R
> dyn.load("RExtension.so")
> .C("presentValue", n=as.integer(4), r=as.double(0.05), t=as.
integer(c(1,2,3,4)), p=as.double(c(3,4,5,6)), res=as.double(0))
 value 2.85714
 value 3.62812
 value 4.31919
 value 4.93621
$n
[1] 4

$r
[1] 0.05

$t
[1] 1 2 3 4

$p
[1] 3 4 5 6

$res
[1] 15.74066

>
```

The desired result is printed as the content of the variable res, which in this case is
15.74066.

Integrating with the Maxima CAS

Implement a class to compute option probabilities that can be accessed using the
Maxima computer algebra system.

Solution

The R environment is an example of a successful application that has been used in the statistical analysis of data sets. Another class of mathematical applications that is commonly employed in data analysis is a computer algebra system (CAS). Financial analysts use such applications to perform algebraic transformations on mathematical functions and expressions. For example, such systems can be used to perform tasks such as solving equations, finding derivatives and integrals, or factoring polynomials. Well-known applications in this category include Mathematica, Maple, and Maxima.

In this section, you will learn how to interact with Maxima, an open source CAS that can be used to assist in the development of mathematical models in finance. You will also understand how to incorporate new or existing C++ code into Maxima, so that you can run iterative experiments with the code while using the Maxima interpreter for visualization purposes.

The Maxima CAS is an open source application that can be freely downloaded and installed from its Internet repository. The main website for the project is located at `http://maxima.sourceforge.net`. Once installed, Maxima can be run using one of the existing front ends that are installed by default. The most commonly used front end to Maxima is wxMaxima, a cross-platform application available for the Windows, Mac OS X, and Linux operating systems. You can also download wxMaxima for free: the latest version is available on the developer's website at `http://wxmaxima.sourceforge.net/`. Figure 16-2 shows the main window of the wxMaxima application.

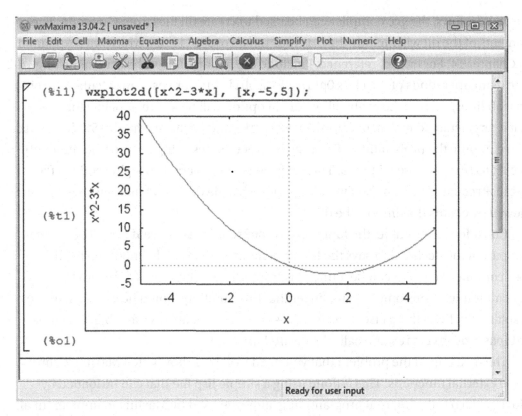

Figure 16-2. *Main window for the wxMaxima application, a front end for the Maxima CAS*

The wxMaxima application is an ideal environment for loading data and performing data analysis, including graphs, summary tables, and simple charts. This functionality can be used to perform quick studies on financial data, based on the results of algorithms such as the ones we discussed in the previous chapters. To integrate existing C++ code with the Maxima environment, you need to create a library that is compatible with the conventions stated on Maxima documentation. In this section, you will learn how to do so with a sample class that calculates options probabilities.

The first step in integrating C++ with Maxima is to create a shared object library that can be loaded by the application. Creating such a library can be easily done with most compilers and integrated development environments (IDEs). I will show how this can be done in Windows with the MingW gcc compiler. Other environments have similar features and most of these instructions will be similar, but you need to check the documentation in Maxima's website.

In Windows, the gcc compiler can be used to create dll files from C++ code. The contents of the dll will include the class OptionsProbabilities, which was explained in Chapter 14. For ease of reference, I include the header file for the class in Listing 16-2. The main operations of the class OptionsProbabilities are used to calculate specific probabilities, such as the probability that an option will be above or below the strike price, as calculated with probFinishAboveStrike and probFinishBelowStrike. You can also compute the probability of finishing between prices using the member function probFinalPriceBetweenPrices. Then, it is necessary to create the glue code in file OptionProbabilityExportedFunctions.cpp. This file declares functions that can be viewed by clients that import the dll.

Consider, for example, the function optionProbFinishAboveStrike. The extern "C" part of the declaration says the function name should not be modified by the C++ compiler, so that it can be found at runtime. The __declspec(dllexport) declaration tells the compiler and linker that this function should be exported in the resulting dll. Everything else is normal C++ code that instantiates an object of class OptionsProbabilities and calls the desired function.

The next part of the problem that you need to solve is how to tell Maxima to find these external functions. This is done using a simple lisp file that can be loaded by Maxima. Lisp is the internal programming language used by Maxima to implement all its functionality. To extend Maxima using lower-level code, you frequently have to create some lisp functions. In this case, however, we will use only two lisp functions that create all the necessary C++ code connecting to the dll created earlier.

The lisp file is named optionProbabilities.l and is shown in Listing 16-2. There are two parts in this file: the first part is a clines function that contains C code that will be compiled and used by Maxima. The second part is a set of lisp declarations for the desired functions. The code inside clines has to be between quotes, and for this reason, it needs to escape any quotes (and backslash characters) using a backslash. Other than that, you can type any normal C statement. You will find that there are four functions that load the desired code from the dll.

The first function in optionProbabilities.l, loadLibrary, is responsible for loading the dll if this has not been done already. This is done using two Windows API functions: LoadLibraryA and GetProcAddress. The function LoadLibraryA takes the name of the library as parameter and returns a reference to it, if the load was successful. The function GetProcAddress, on the other hand, retrieves a pointer to the function named as in its second parameter. Once loadLibrary has completed its work, you will have pointers to the three functions exported in the file optprob.dll.

The next three functions in optionProbabilities.l are then used to call each of the desired functions in the dll. The file ends with three declarations that tell Maxima to accept these functions as top-level operations, along with the desired types.

Complete Code

In Listing 16-2, you can find the complete code for using the OptionsProbabilities class from the Maxima CAS. After the C++ code, you can also see the lines of lisp code necessary to import the class into the Maxima environment.

Listing 16-2. Class OptionsProbabilities and Associated Maxima Code

```
//
// OptionsProbabilities.h

#ifndef __FinancialSamples__OptionsProbabilities__
#define __FinancialSamples__OptionsProbabilities__

#include <vector>

class OptionsProbabilities {
public:
    OptionsProbabilities(double initialPrice, double strike, double
    avgStep, int nDays);
    OptionsProbabilities(const OptionsProbabilities &p);
    ~OptionsProbabilities();
    OptionsProbabilities &operator=(const OptionsProbabilities &p);

    void setNumIterations(int n);

    double probFinishAboveStrike();
    double probFinishBelowStrike();
    double probFinalPriceBetweenPrices(double lowPrice, double highPrice);
    std::vector<double> getWalk();
```

```cpp
private:
    double m_initialPrice;
    double m_strike;
    double m_avgStep;
    int m_numDays;
    int m_numIterations;

    double gaussianValue(double mean, double sigma);
    double getLastPriceOfWalk();
};

#endif /* defined(__FinancialSamples__OptionsProbabilities__) */

//
// OptionProbabilityExportedFunctions.cpp

#include "OptionsProbabilities.h"

extern "C" double __declspec(dllexport) optionProbFinishAboveStri
ke(double initialPrice, double strike, double avgStep, int nDays)
{ OptionsProbabilities optP(initialPrice, strike, avgStep, nDays);
 return optP.probFinishAboveStrike();
}

extern "C" double __declspec(dllexport) optionProbFinishBelowStri
ke(double initialPrice, double strike, double avgStep, int nDays)
{ OptionsProbabilities optP(initialPrice, strike, avgStep, nDays);
return optP.probFinishBelowStrike();
}

extern "C" double __declspec(dllexport) optionProbFinishBetweenPrices
(double initialPrice, double strike, double avgStep, int nDays, double
lowPrice, double highPrice) { OptionsProbabilities optP(initialPrice,
strike, avgStep, nDays); return optP.probFinalPriceBetweenPrices(lowPrice,
highPrice);
}
```

```
;;
;; file  optionProbabilities.l
;;
(lisp:clines "
static double (*optionProbFinishAboveStrike_)(double,double,double,int) =
NULL;
static double (*optionProbFinishBelowStrike_)(double,double,double,int) =
NULL;
  static double (*optionProbFinishBetweenPrices_)(double,double,double,int,
  double,double) = NULL;
__declspec (dllimport) void *__stdcall LoadLibraryA(const char *);
void *__stdcall GetProcAddress(void *,const char *);
__declspec (dllimport) unsigned int __stdcall GetLastError(void);

static int libraryLoaded = 0;
static const char *libName = \"optprob.dll\";

static int loadLibrary() {
   void *lib = LoadLibraryA(libName);
   if (!lib) return 0;
   optionProbFinishAboveStrike_ = GetProcAddress(lib,
   \"optionProbFinishAboveStrike\");
   optionProbFinishBelowStrike_ = GetProcAddress(lib,
   \"optionProbFinishBellowStrike\");
   optionProbFinishBetweenPrices_ = GetProcAddress(lib,
   \"optionProbFinishBetweenPrices\");
   libraryLoaded = 1;
   return 1;
}

double l_optionProbFinishAboveStrike(double a,double b,double c,int d) {
   if (!libraryLoaded && !loadLibrary()) return -1; /* error code */
   if (!optionProbFinishAboveStrike_) return -2;
   return optionProbFinishAboveStrike_(a, b, c, d);

}
```

```
double l_optionProbFinishBelowStrike(double a,double b,double c,int d) {
    if (!libraryLoaded && !loadLibrary()) return -1; /* error code */
    return optionProbFinishAboveStrike_(a, b, c, d);
}
double l_optionProbFinishBetweenPrices(double a,double b,double c,int
d,double e,double f) {
    if (!libraryLoaded && !loadLibrary()) return -1; /* error code */
    return optionProbFinishBetweenPrices_(a, b, c, d, e, f);
}
")

(lisp:defentry $optionProbFinishAboveStrike  (lisp:double lisp:double
lisp:double lisp:int)
    (lisp:double "l_optionProbFinishAboveStrike"))

(lisp:defentry $optionProbFinishBelowStrike (lisp:double lisp:double
lisp:double lisp:int)
    (lisp:double "l_optionProbFinishBelowStrike"))

(lisp:defentry $optionProbFinishBetweenPrices(lisp:double lisp:double
lisp:double lisp:int lisp:double lisp:double)
    (lisp:double "l_optionProbFinishBetweenPrices"))
```

Running the Code

Once you have created the code described in Listing 16-2, the next step is to build and run it using the Maxima CAS. First, you need to create the dll using the MingW gcc compiler. Here is the command line I used (you may need to adjust this to use the library paths in your system).

```
g++ "-IC:\\bin\\boost_1_55_0\\" -O0 -g3 -Wall -c -fmessage-length=0 -o \
"src\\OptionProbabilityExportedFunctions.o" "..\\src\\
OptionProbabilityExportedFunctions.cpp"
g++ "-IC:\\bin\\boost_1_55_0\\" -O0 -g3 -Wall -c -fmessage-length=0 -o \
"src\\OptionsProbabilities.o" "..\\src\\OptionsProbabilities.cpp"
g++ -shared -o optprob.dll "src\\TestClass.o" "src\\OptionsProbabilities.o"
\ "src\\OptionProbabilityExportedFunctions.o"
```

Once the dll has been created, you can now use Maxima to load it. For these instructions to work, you need to make sure that Maxima is using GCL (Gnu Common Lisp) as the underlying lisp engine (you can determine this when downloading or building Maxima). The following lines tell you this information when Maxima is started:

```
Maxima 5.31.2 http://maxima.sourceforge.net
using Lisp GNU Common Lisp (GCL) GCL 2.6.8 (a.k.a. GCL)
Distributed under the GNU Public License. See the file COPYING.
```

The following is a transcript of a session with Maxima that shows how this works:

```
/* [wxMaxima: input   start ] */
(%i28) :lisp (compile-file
  "c:/MaximaCode/optionProbabilities.l" :c-file t :h-file t)

(%o28) Compiling c:/MaximaCode/optionProbabilities.l.
End of Pass 1.
End of Pass 2.
OPTIMIZE levels: Safety=2, Space=3, Speed=3
Finished compiling c:/MaximaCode/optionProbabilities.l.
#pc:/MaximaCode/optionProbabilities.o

(%i29) :lisp (load "c:/MaximaCode/optionProbabilities.o");
(%o29) OK

(%i30) optionprobfinishbellowstrike(30.0, 35.0, 0.01,  800);
(%o30)                              0.246
```

In this example, the first command starting with :lisp is used to compile the lisp file displayed. Once this is done, the combination of lisp and C code is saved as an object file, which is then named optionProbabilities.o. The second command starting with :lisp is used to load the resulting object file into the system. After this is finished, the functions typed in optionProbabilities.l will become available to Maxima.

The last step shows how to invoke the desired functions. For example, you can call the function optionprobfinishbelowstrike with a set of parameters that define the price of the underlying, the strike, the volatility, and the number of time periods.

Conclusion

This chapter contains programming examples that show how to integrate existing C++ classes with two open source mathematical applications: R and Maxima. The number of users for environments like these is growing due to the ease of programming with such a dynamic language, along with the ability to see immediate results from computations. Although these mathematical applications already contain a lot of functionality, most financial developers also need to access existing C++ code when making more complex analyses.

The examples in this chapter show you how to interface your C++ code with these two popular open source mathematical applications. First, you learned how C++ integration can be achieved with the R programming environment. You have seen an example where a set of payments are received as input, and the code calculates the present value of these payments.

The second C++ example shows how to export an existing class to the Maxima CAS. With Maxima, you have access to a large number of mathematical tools to analyze and display data. The process of exporting C++ libraries to Maxima involves the creation of a dll and the use of some glue code written in C. Once you have these tools, it is possible to access any C++ code from Maxima.

One of the secrets of creating useful financial software is to employ computational resources as efficiently as possible. In this way, it is possible to learn even more from existing investment data while making faster and more accurate decisions. In the next chapter, you will learn about multithreading, a programming technique that is frequently used to improve the performance of numerical and networking code used in financial applications. Because of the growing use of multicore processors in servers and even desktop machines, the use of multithreading has become a necessity for modern code written in C++. You can access such multiprocessing features from C++ using a few standard libraries, as you will see in Chapter 17.

CHAPTER 17

Multithreading

C++20 applications are used in contexts where computational performance has great importance. The need for performance is even more prominent in financial applications such as high-frequency trading, where the difference between profit and loss may be just a few microseconds away. In such cases where performance is a requirement, it is very important to take advantage of the resources made available in modern CPUs (central processing units). In particular, multicore processing is one of the main features provided by new CPUs, with the number of available cores constantly increasing along with the complexity of chips.

To benefit from multicore processors, however, C++ programmers need to learn a few parallel programming techniques that have rapidly become part of the C++ repertoire during the last decades. Using multiple processes is a possible way to explore this computational power. Multithreading, a method used to run several concurrent tasks inside the same process, is another technique that has the potential to take advantage of two or more cores at the same time.

In this chapter, you will see a few examples that explore multithreading strategies for C++ programmers. With the knowledge provided in this chapter, you will be able to take full advantage of existing multicore systems on your applications. While multithreading is a useful strategy to employ in today's applications, it will become even more important in the future, as desktop and server manufacturers are expected to continue to add more cores to their processors.

In the traditional approach for multithreading, C++ programmers used libraries created to facilitate the access to the multithreading facilities provided by the operating system. A popular example is the pthreads library. However, the current standard C++20 provides another approach, whereby the use of threading is directly supported by the language, through templates in the <thread> header file. In this chapter, you will learn both ways. This is important because much of the existing multithreading code uses nonstandard libraries. However, new code should preferably use the templates provided by the standard library.

© Carlos Oliveira 2021
C. Oliveira, *Practical C++20 Financial Programming*, https://doi.org/10.1007/978-1-4842-6834-6_17

The following are some of the topics discussed in this chapter:

- Using the pthreads library: pthreads is a standard library that can be used to create and maintain multiple threads in the same process. In a multicore machine, creating threads is one of the most common ways to explore the parallel abilities of current architectures. You will learn how to create applications that employ the pthreads library to achieve parallel computation.

- Running algorithms in parallel threads: You will see how to decompose problems in separate threads and combine their results into a new solution. As an example, I present a modified parallel algorithm for the calculation of options probabilities.

- Thread synchronization: The use of multiple threads introduces the problem of synchronizing resources. You will learn how to use synchronization primitives such as mutexes to guarantee that resources are accessed and modified by only one thread at each time.

- STL threads: The new multithreading classes and templates provided by the latest releases of the C++ standard.

Creating Threads with the Pthreads Library

Create a C++ class that distributes its work through several processing threads using the standard pthreads library.

Solution

Multithreading is one of the software solutions that have been created to support parallel computation. A thread is a unit of processing that can be performed in parallel along with other parts of a program, so that two or more segments of a program can be executed concurrently. In a multicore machine, this means that the same program may efficiently use two or more cores to perform additional work. Depending on how the code is organized, the careful use of multithreading techniques provides a good opportunity to improve the throughput of the whole algorithm.

To use multithreading, however, support from the operating system is necessary. Since each operating system implements multithreading internally in its own way, it used to be the case that a multithreading application would be dependent on the operating system and architecture used. To avoid this problem, a standard pthreads (POSIX threads) library was proposed and adopted as part of POSIX. The pthreads library is available for multiple operating systems, including UNIX, Mac OS X, and even Windows (e.g., you can use the Cygwin libraries to access pthreads on Windows). Table 17-1 provides a quick summary of functions in the pthreads API (application programming interface) that are available for application developers.

Table 17-1. *List of Commonly Used Functions in the Pthreads Library*

Function name	Description
pthread_create	Creates a new thread
pthread_exit	Finishes an existing thread
pthread_join	Joins an existing thread, returning only after the thread exits
pthread_detach	Detaches from a thread
pthread_attr_init	Initializes an attribute data structure
pthread_attr_destroy	Destroys an attribute data structure
pthread_attr_setstacksize	Sets the size of the stack for a new thread
pthread_cancel	Cancels the thread execution
pthread_mutex_init	Initializes a mutex synchronization primitive
pthread_mutex_destroy	Destroys a mutex
pthread_mutex_lock	Locks a mutex
pthread_mutex_unlock	Unlocks an existing mutex
sem_init	Initializes a semaphore synchronization primitive
sem_destroy	Destroys a semaphore
sem_wait	Waits on a semaphore

While the pthreads library is written for compatibility with pure C programs, it can easily be used as part of C++ applications. It is simple to create a wrapper around threads created with pthreads so that they can be more easily accessible from C++ code. In this section, you will see a C++ code example for creating a simple multithreaded application using pthreads. The techniques used as well as the general concept of thread creation and synchronization can be used in your own programs.

To create a separate thread of execution within a program, you need to use the pthread_create function. It receives as parameters an identifier (integer value), a pointer to an attribute structure (which can be null if not used), a pointer to a function that will be executed by the thread, and a pointer to the arguments to the thread function. The function returns zero if no error happened; otherwise, it returns an integer error identifier.

After the pthread_create function is executed, the program starts another thread from the specified function. That thread is independent of the original program and may run in the same or in a separate core, if there is one available in the host machine as determined by the operating system's thread scheduler. A thread can be terminated either by reaching the end of the thread function or by explicitly calling the pthread_exit function.

In this section, I show how to access the functionality provided by the pthreads library from a C++ class. For this purpose, I introduce the Thread class, which encapsulates the concept of a running thread. The goal of this class is to become a base class for concrete thread classes. The only member function that is required in each new subclass of Thread is run(), which determines the code that will be executed by the new thread.

Notable methods in the Thread class are the following:

- start: Needs to be called to start the execution of the thread.

- endThread: Can be called to terminate the current thread.

- setJoinable: Determines if the thread can be joined by other threads.

- join: Allows a caller to join this thread, in such a way that the caller will continue its execution only after the thread has terminated.

- run: This member function needs to be implemented in each subclass and defines the code that will run in its own thread.

The Thread class uses a C function called thread_function and is defined in the Thread.cpp file:

```
void *thread_function(void *data)
{
    Thread *t = reinterpret_cast<Thread*>(data);
    t->run();
    return nullptr;
}
```

The signature of this function is determined in the pthreads library. The function is called as soon as a new thread is created. The main idea is that the data pointer passed to the function is in fact a pointer to a Thread object. Once it is retrieved using the reinterpret_cast operator, the object can be used to perform the run member function. Depending on the concrete subclass of Thread, the run method may do any task desired by the creator of the subclass. This is enough to guarantee that the code will run as a parallel thread.

Note   Remember that the reinterpret_cast operator can be used to convert between any two types in C++. Therefore, it is important to be careful when using this operator, since there is no type checking performed by the compiler once it is applied.

Other than that, the start() and endThread() functions use the corresponding pthread API functions to perform the creation of a new thread and to exit from an existing thread, respectively. This is how these functions are implemented:

```
void Thread::start()
{
    pthread_create(&m_data->m_thread, &m_data->m_attr, thread_function,
    this);
}

void Thread::endThread()
{
    pthread_exit(nullptr);
}
```

Complete Code

You can find the complete implementation for the Thread class in Listing 17-1.

Listing 17-1. Thread Class and a Sample Implementation

```
//
//   Thread.h

#ifndef __FinancialSamples__Thread__
#define __FinancialSamples__Thread__

struct ThreadData;

class Thread {
public:
    Thread();
    virtual ~Thread();
private:
    Thread(const Thread &p); // no copy allowed
    Thread &operator=(const Thread &p); // no assignment allowed

public:
    virtual void run() = 0;
    void start();
    void endThread();
    void setJoinable(bool yes);
    void join();

private:
    ThreadData *m_data;
    bool m_joinable;
};

#endif /* defined(__FinancialSamples__Thread__) */

//
//   Thread.cpp

#include "Thread.h"
```

```cpp
#include <pthread.h>
#include <iostream>

using std::cout;
using std::endl;

struct ThreadData {
    pthread_t m_thread;
    pthread_attr_t m_attr;
};

namespace {

    void *thread_function(void *data)
    {
        Thread *t = reinterpret_cast<Thread*>(data);
        t->run();
        return nullptr;
    }
}

Thread::Thread()
: m_data(new ThreadData),
  m_joinable(false)
{
    pthread_attr_init(&m_data->m_attr);
}

Thread::~Thread()
{
    if (m_data)
    {
        delete m_data;
    }
}
```

```cpp
void Thread::start()
{
    pthread_create(&m_data->m_thread, &m_data->m_attr, thread_function, this);
}

void Thread::endThread()
{
    pthread_exit(nullptr);
}

void Thread::setJoinable(bool yes)
{
    pthread_attr_setdetachstate(&m_data->m_attr, yes ? PTHREAD_CREATE_
    JOINABLE : PTHREAD_CREATE_DETACHED);
    m_joinable = yes;
}

void Thread::run()
{
    cout << " no concrete implementation provided " << endl;
}

void Thread::join()
{
    if (!m_joinable)
    {
        cout << " thread cannot be joined " << endl;
    }
    else
    {
        void *result;
        pthread_join(m_data->m_thread, &result);
    }
}

// --- sample implementation
```

```
class TestThread : public Thread {
public:
    virtual void run();
};

void TestThread::run()
{
    cout << " this is a test implementation " << endl;
    endThread();
}

int main()
{
    Thread *myThread = new TestThread;
    myThread->setJoinable(true);
    myThread->start();
    myThread->join();
    return 0;
}
```

Running the Code

The code displayed in Listing 17-1 can be built using any standards-compliant compiler, such as gcc, llvm, or Visual Studio. Just remember to add a link line including the pthreads library. The following is a command line used to build the sample application (tested on Mac OS X):

```
$ gcc -o threadTest Thread.cpp -lpthreads
$ ./threadTest
  this is a test implementation
```

Calculating Options Probabilities in Parallel

Create a multiprocessing version of the class that calculates options probabilities. Use the pthreads library to distribute work among several threads.

Solution

The use of parallel processing techniques is highly indicated for problems that require massive amounts of computation. This is especially true when the problem can be easily decomposed, in which case it becomes a matter of distributing the right amount of work to each thread and waiting for the results.

A good example of such a process is a Monte Carlo-based algorithm. The simulation process can run in any number of threads, and their results can be combined easily into a single number. This is the case, for example, of calculating options probabilities. As you saw in Chapter 14, Monte Carlo techniques are effective for the simulation of options probabilities. At each step, you just need to simulate a new random walk and use that information to improve the current estimate of the probability.

To adapt the Monte Carlo algorithm to the determination of options probabilities, the first step is to correctly define the way in which the problem will be decomposed. This is easy to do here, because each loop of the computation is independent of the other. In this case, you can do this by telling each thread to run a certain number of iterations of the Monte Carlo method. At the end, you can combine the results found by each thread and calculate the final result as the average of the values returned by all threads.

The algorithm just described is implemented in the `ParallelOptionsProbabilities` class. The class is an outer shell that invokes several threads to run the desired algorithm. The real work is done in a class derived from `Thread`, and called `RandomWalkThread`. As any other subclass of `Thread`, it needs to implement the `run()` member function, which is called from the separate thread. Inside `RandomWalkThread`, you will find a member variable, `m_result`, which stores the output of the Monte Carlo process. After the thread is finished, this member variable can be used to retrieve the final value of the computation.

The `run` member function is very similar to the code you already saw in the `OptionsProbabilities` class. The main difference is that the output is stored in the `m_result` member variable. The work of `RandomWalkThread` objects is orchestrated inside the `ParallelOptionsProbabilities` class. The important member function for the job is `probFinishAboveStrike`.

```
double ParallelOptionsProbabilities::probFinishAboveStrike()
{
    const int numThreads = 20;

    vector<RandomWalkThread*> threads(numThreads);
```

```
for (int i=0; i<numThreads; ++i)
{
    threads[i] = new RandomWalkThread(m_numSteps, m_step,
    m_strikePrice);
    threads[i]->setJoinable(true);
    threads[i]->start();
}

for (int i=0; i<numThreads; ++i)
{
    threads[i]->join();
}

double nAbove = 0;
for (int i=0; i<numThreads; ++i)
{
    nAbove += threads[i]->result();
    delete threads[i];
}

return nAbove/(double)(numThreads);
}
```

At the beginning of the member function, several threads are created and added to the threads vector. You need to define these threads as joinable, so that it is possible to wait on the result of each thread. The next step is to start the threads so that each of them can perform the desired computations. Then, the second loop is used to join the already created threads. By doing this, the main thread can wait while the computation is being performed in parallel. When all threads are finished, the main thread will be resumed as a result of the call to join(). Finally, you can store the data returned by each thread using the result() member function. The thread objects may be deleted at this time to avoid memory leaks. In the last line of the probFinishAboveStrike member function, you can see how the calculated data can be combined. In this case, it is enough to return the sum of values above the strike prices and divide that value by the number of threads used.

Complete Code

Listing 17-2 displays the ParallelRandomWalk class. There is a sample main() function that can be used for testing, as can be seen at the end of the listing.

Listing 17-2. Class ParallelRandomWalk

```
//
//
//  ParallelRandomWalk.h

#ifndef __FinancialSamples__ParallelRandomWalk__
#define __FinancialSamples__ParallelRandomWalk__

class ParallelOptionsProbabilities  {
public:
    ParallelOptionsProbabilities(int size, double strike, double sigma);
    ParallelOptionsProbabilities(const ParallelOptionsProbabilities &p);
    ~ParallelOptionsProbabilities();
    ParallelOptionsProbabilities &operator=(const
    ParallelOptionsProbabilities &p);

    double probFinishAboveStrike();

private:
    int m_numSteps;        // number of steps
    double m_step;         // size of each step (in percentage)
    double m_strikePrice; // starting price
};

#endif /* defined(__FinancialSamples__ParallelRandomWalk__) */

//
// ParallelOptionsProbabilities.cpp

#include "ParallelOptionsProbabilities.h"

#include "Thread.h"

#include <pthread.h>
#include <cstdlib>
```

```
#include <vector>

#include <boost/random/normal_distribution.hpp>
#include <boost/random.hpp>

using std::vector;
using std::cout;
using std::endl;

static boost::rand48 random_generator;

using std::vector;

/// ---

class RandomWalkThread : public Thread {
public:
    RandomWalkThread(int num_steps, double sigma, double startPrice);
    ~RandomWalkThread();
    virtual void run();

    double gaussianValue(double mean, double sigma);
    double getLastPriceOfWalk();

    double result();
private:
    int m_numberOfSteps;      // number of steps
    double m_sigma;           // size of each step (in percentage)
    double m_startingPrice;   // starting price

    double m_result;
};

RandomWalkThread::RandomWalkThread(int numSteps, double sigma, double
startingPrice)
: m_numberOfSteps(numSteps),
  m_sigma(sigma),
  m_startingPrice(startingPrice)
{
}
```

```
RandomWalkThread::~RandomWalkThread()
{
}

double RandomWalkThread::gaussianValue(double mean, double sigma)
{
    boost::random::normal_distribution<> distrib(mean, sigma);
    return distrib(random_generator);
}

double RandomWalkThread::result()
{
    return m_result;
}

double RandomWalkThread::getLastPriceOfWalk()
{
    double prev = m_startingPrice;

    for (int i=0; i<m_numberOfSteps; ++i)
    {
        double stepSize = gaussianValue(0, m_sigma);
        int r = rand() % 2;
        double val = prev;
        if (r == 0) val += (stepSize * val);
        else val -= (stepSize * val);
        prev = val;
    }
    return prev;
}

void RandomWalkThread::run()
{
    cout << " running thread " << endl;

    int nAbove = 0;
    for (int i=0; i<m_numberOfSteps; ++i)
```

```
    {
        double val = getLastPriceOfWalk();
        if (val >= m_startingPrice)
        {
            nAbove++;
        }
    }

    m_result = nAbove/(double)m_numberOfSteps;
}

// ---

ParallelOptionsProbabilities::ParallelOptionsProbabilities(int size, double
start, double step)
: m_numSteps(size),
  m_strikePrice(start),
  m_step(step)
{
}

ParallelOptionsProbabilities::ParallelOptionsProbabilities(const
ParallelOptionsProbabilities &p)
: m_numSteps(p.m_numSteps),
  m_strikePrice(p.m_strikePrice),
  m_step(p.m_step)
{
}

ParallelOptionsProbabilities::~ParallelOptionsProbabilities()
{
}

ParallelOptionsProbabilities &ParallelOptionsProbabilities::operator=(const
ParallelOptionsProbabilities &p)
{
```

```cpp
    if (this != &p)
    {
        m_numSteps = p.m_numSteps;
        m_strikePrice = p.m_strikePrice;
        m_step = p.m_step;
    }
    return *this;
}

double ParallelOptionsProbabilities::probFinishAboveStrike()
{
    const int numThreads = 20;

    vector<RandomWalkThread*> threads(numThreads);
    for (int i=0; i<numThreads; ++i)
    {
        threads[i] = new RandomWalkThread(m_numSteps, m_step,
        m_strikePrice);
        threads[i]->setJoinable(true);
        threads[i]->start();
    }

    for (int i=0; i<numThreads; ++i)
    {
        threads[i]->join();
    }

    double nAbove = 0;
    for (int i=0; i<numThreads; ++i)
    {
        nAbove += threads[i]->result();
        delete threads[i];
    }

    return nAbove/(double)(numThreads);
}
```

```
int main()
{
    ParallelOptionsProbabilities rw(100, 50.0, 52.0);
    double r = rw.probFinishAboveStrike();
    cout << " result is " << r << endl;
    return 0;
}
```

Running the Code

I have compiled and executed the code displayed in Listing 17-2 using the gcc compiler on a Mac OS X machine. Any standards-compliant compiler can be used for this purpose. The following is a sample of the expected output:

```
./parallelOptProb
 running thread
 running thread
 ...
 running thread
result is 0.487
```

Using Mutexes to Prevent Unsynchronized Access

In this section, we will write a C++ class that implements a parallel algorithm where mutexes are used to synchronize shared data.

Solution

Multithreading is a convenient way to distribute computational work into two or more processor cores, which can lead to an increase in performance for the whole application. However, while multithreading has numerous advantages, it also adds to the complexity of the software design. For example, one of the problems that need to be solved in multithreading architectures is the access to resources shared between threads. If a variable in memory is used in two or more threads, its access needs to be synchronized so that separate threads will not try to change values concurrently, for example.

Once a new thread has been created, it is necessary to manage it, using mechanisms provided by the pthread library. In the simplest case, the new thread is independent and does not need to be synchronized with the original (parent) thread. More commonly, however, it is necessary to perform synchronization between separate threads that use the same data. The greater the need for synchronization, the greater the amount of work spent on managing the shared data.

A section of code where two or more threads can access a shared resource is called a *critical section*. The critical sections are the areas of the code where shared resources need to be protected, in order to avoid conflicts.

To avoid the conflicts inherent to the existence of critical sections, most multithreading APIs provide primitives that can be used to implement synchronized operations. Such operations have proved effective in enabling resource sharing between threads. There are a number of such primitives, such as semaphores, mutexes, and messages, among others. The pthreads library provides direct support for some of the most common of such mechanisms, including mutexes, which can be used to guarantee that only one thread is able to access a particular critical section.

A *mutex* is a synchronization mechanism that can be used to coordinate the work of two or more threads. The mutex state is used to determine if a thread has permission to operate on a particular resource, such as a variable in memory. When a thread tries to access the value of the mutex, two things can happen: if the mutex state indicates that the critical section is available, then the thread can directly proceed to the critical section. However, if the mutex state indicates a busy state, the thread making the request stops its execution and is sent to a waiting area created by the operating system. Operation will resume only when the resource has been made available by other threads. All this waiting and resuming activity is coordinated by the operating system.

Mutexes are implemented in the pthreads API and have the type `thread_mutex_t`. A new mutex can be created using the function `pthread_mutex_init` and destroyed using the function `pthread_mutex_destroy`.

A mutex needs to be acquired and locked when a shared resource is about to be used. This guarantees that the mutex will be available for only one thread at a time. This is done using the function `pthread_mutex_lock`. This function will automatically interrupt the thread if the mutex is not available and force the thread to wait until the mutex has been released. You can also try to access the mutex without a forced wait using `pthread_mutex_trylock`. This will return an error code in case the mutex is currently not available, and you will be free to try it later. Finally, once a mutex has been acquired, you need to unlock the mutex at the end of the synchronized operation. This is necessary to make sure that other threads can enter the critical section and use the recently released resource. You can unlock a mutex using the function pthread_mutex_unlock.

While theoretical proof of the effectiveness of the mutex can be complex, its use is very simple. In the coding example in Listing 17-3, you will have a class called Mutex that encapsulates the concept of a mutex synchronization operation. There are two main functions provided by this class: lock and unlock. The first member function is called at the beginning of a critical section. The second important member function in the class is unlock, which should be called at the end of a critical section. The Mutex class is also responsible for initializing the pthread mutex at the constructor with pthread_mutex_init and destroying it at the destructor with pthread_mutex_destroy.

The second class used to encapsulate the mutex concept is MutexAccess. This class is responsible for guaranteeing that each access to the mutex is composed of a pair of calls to the lock() and unlock() member functions of Mutex. The lock() member function is directly called in the constructor, and unlock() is automatically called in the destructor of MutexAccess. Therefore, if the critical section ends right at the end of the scope where the MutexAccess object is declared, you don't need to worry about unlocking it, since the RAII idiom guarantees that the mutex will be automatically unlocked when the destructor is called.

In the MutexTestThread, we have an example of using the Mutex class inside a thread. The task demonstrated is really simple, but it illustrates how the mutex can be used to provide synchronization of access to shared resources. Here, the shared resource is the variable result, of double type. This variable is used to hold the desired calculation; however, it is being accessed in all threads in the application. In order to synchronize access to this variable, you need to use a mutex. An object of the class MutexAccess can be instantiated, resulting in the mutex (named m_globalMutex) being locked. After the lock has been acquired, you can now safely check the value and make changes to the reference variable. Finally, at the end of the run() member function, the lock will be released automatically.

Complete Code

You can view the complete code for the Mutex and MutexAccess classes in Listing 17-3. An example of their use is also shown in class MutexTestThread.

Listing 17-3. The Mutex Class

```
//
//  Mutex.h

#ifndef __FinancialSamples__Mutex__
#define __FinancialSamples__Mutex__

struct MutexData;

class Mutex {
public:
    Mutex();
    ~Mutex();

    void lock();
    void unlock();
private:
    Mutex(const Mutex &p);  // copy not allowed
    Mutex &operator=(const Mutex &p);  // assignment not allowed

    MutexData *m_data;
};

class MutexAccess {
public:
    MutexAccess(Mutex &m);
    ~MutexAccess();
private:
    MutexAccess &operator=(const MutexAccess &p);
    MutexAccess(const MutexAccess &p);

    Mutex &m_mutex;
};

#endif /* defined(__FinancialSamples__Mutex__) */

//
//  Mutex.cpp
```

```cpp
#include "Mutex.h"

#include "Thread.h"

#include <pthread.h>
#include <cstdlib>
#include <vector>
#include <iostream>

using std::vector;
using std::cout;
using std::endl;

struct MutexData {
    pthread_mutex_t m_mutex;
};

Mutex::Mutex()
: m_data(new MutexData)
{
    pthread_mutex_init(&m_data->m_mutex, NULL);
}

Mutex::~Mutex()
{
    if (m_data)
    {
        pthread_mutex_destroy(&m_data->m_mutex);
        delete m_data;
    }
}

void Mutex::lock()
{
    pthread_mutex_lock(&m_data->m_mutex);
}

void Mutex::unlock()
```

```
{
    pthread_mutex_unlock(&m_data->m_mutex);
}
/// ----

MutexAccess::MutexAccess(Mutex &m)
: m_mutex(m)
{
    m_mutex.lock();
}

MutexAccess::~MutexAccess()
{
    m_mutex.unlock();
}
/// ----

class MutexTestThread  : public Thread {
public:
    MutexTestThread(double &result, double incVal);
    ~MutexTestThread();

    void run();
private:
    double &m_result;
    double m_incValue;

    static Mutex m_globalMutex;
};

Mutex MutexTestThread::m_globalMutex;  // global mutex is static

MutexTestThread::MutexTestThread(double &result, double incVal)
: m_result(result),
  m_incValue(incVal)
```

```
{
}

MutexTestThread::~MutexTestThread()
{
}

void MutexTestThread::run()
{
    MutexAccess maccess(m_globalMutex);   // mutex is locked here
    cout << " accessing data " << endl; cout.flush();
    if (m_result > m_incValue)
    {
        m_result -= m_incValue;
    }
    else
    {
        m_incValue += m_incValue;
    }

    // mutex is automatically unlocked
}

int main()
{
    int nThreads = 10;

    vector<Thread*> threads(nThreads);
    double price = rand() % 25;

    for (int i=0; i<nThreads; ++i)
    {
        threads[i] = new MutexTestThread(price, (double)(rand() % 10));
        threads[i]->setJoinable(true);
        threads[i]->start();
    }
    for (int i=0; i<nThreads; ++i)
```

```
    {
        threads[i]->join();
    }

    cout << " final price is " << price << endl;
    return 0;
}
```

Running the Code

You can compile this code using any standard C++ compiler. I performed the test on a machine running the Mac OS X operating system. The following is a display of sample results:

```
accessing data
accessing data    ...
accessing data
accessing data
accessing data
final price is 2
```

Creating Threads Using the Standard Library

In the previous section, you learned how to create multithreaded programs using the pthreads library. In C++20, it is also possible to create multithreaded code using the standard library. The support is provided through the <thread> header file.

To make simple multithreaded programs using the STL, it is not necessary to create new classes or objects. The class std::thread already has the ability to perform multithreaded operations using as input a function, a lambda, or a functional object.

Consider the following example:

```
#include <thread>
#include <iostream>
#include <vector>
```

```cpp
void compute_max(const std::vector<double> &values)
{
    auto total = 0.0;
    for (auto v : values)
    {
        total += v;
    }
    std::cout << " total: " << total << std::endl;
}

int main()
{
    std::vector<double> v = {0, 5, 3, 2, 5, 3};
    std::thread first_tread(compute_max, v);
    first_tread.join();
    return 0;
}
```

Here, we define a simple function called compute_max, which receives as parameter a vector of double numbers. This function could be any type of operation that takes a long time and that we would like to move to a separate thread. To create a new thread using this function, we just need to use the std::thread class in the <thread> header file.

The std::thread class takes as parameters the name of the function (or functional object) you want to use, along with zero or more parameters that will be passed to that function. In the previous example, we have the vector named v as the single parameter. This could be expanded to other parameters if required by the function compute_max.

Finally, the first_thread object calls the join() method, to indicate that the main function will join the execution of that thread, until it is complete. If we didn't want to stop until the thread is completed, we could have used instead the detach() method, which allows the thread to run independently while the current function continues its operation.

Conclusion

The constant development of multicore processors and architectures has greatly expanded the computational capacity of modern machines. However, to explore such multicore architectures, it is necessary to change the way you program. Modern high-performance programming has an increased focus on multiprocessing techniques, which allow applications to access more than one core and as result improve the performance of the system.

Multithreading is a programming technique that allows more than one thread of execution per process. If the machine has more than one processor, multithreading allows you to access these processing cores to perform additional work. In this chapter, you learned how to create, terminate, and manage threads using the standard pthreads library.

In the first programming example, you learned about the pthreads library and how it can be used to create new threads. You saw how to design a C++ class to encapsulate the pthreads function calls. Using pthreads, you can simplify your multithreading applications, as it abstracts away system-dependent APIs for multithreading.

Next, you learned how to apply pthreads to a common problem on options. You saw that, for some problems, it is easy to distribute the necessary work into separate units of computation. Using C++, you can encapsulate such code segments into different objects.

In the next section, you learned about synchronization primitives and how they are implemented using the pthreads library. I introduced a class that can be used to model the operation of a mutex. You can readily apply the `Mutex` class to other financial programming projects.

Finally, I also explained how the new C++ standard C++20 provides direct support for threads without the use of a separate library like pthreads. Thus, for new code, it is possible to simplify the applications and rely on the STL. While much of the existing multithreading code still uses libraries like pthread, it is important to learn how to do this using the standard and use it in new projects.

With this chapter, I have completed a general presentation of technical tools used to create high-performance financial applications in C++. I hope you have enjoyed learning about the features of C++ and how they can be applied to the solution of common problems in the financial industry.

APPENDIX A

Features of C++20

C++ is a language in constant evolution. Since its first release in the 1980s, new concepts and techniques that started as research topics became an integral part of the language. The latest revision of the C++ standard is C++20, which is itself a major addition to previous standards such as C++11, C++14, and C++17. These updates to C++ are already part of major compilers, so it is important to understand what these modifications bring for developers.

In this appendix, I will provide a summary of the most important changes introduced in these recent C++ standards. You will learn about the following topics:

- auto-typed variables: A syntax that allows automatic type detection

- Lambdas: Creating functions in place and sharing variables from a local environment

- User-defined literals: Creating literals with user-define behavior

- Range-based `for`: A new form of the `for` loop which simplifies container manipulation

- Rvalue references: A new technique to implement move semantics into user-defined types

- New function declarator syntax: A syntax for function where the return type is automatically detected

- Delegating constructors: How to delegate class initialization to a single constructor

- Inheriting constructors: Directly using constructors defined in a parent class

- Generalized attributes: How to declare attributes for C++ elements using a unified syntax

© Carlos Oliveira 2021
C. Oliveira, *Practical C++20 Financial Programming*, https://doi.org/10.1007/978-1-4842-6834-6

- Generalized constant expressions: Defining expressions that can be used at compilation time by other expression, including templates

- Null pointer constant: A new constant that uniquely defines a null pointer

- Right angle brackets: A simplification of template syntax, avoiding common confusions with the shift operator

- Initializer lists: A general way to perform initialization of C++ variables

Automatic Type Detection

One of the main features of C++ is the use of types to check the program during compilation time. This feature, known as strong type checking, allows programmers to rely on the compiler to find many bugs that would take a lot of time to remove otherwise. It is generally accepted that static checking is a useful feature, especially for large-scale projects, where hundreds or even thousands of classes can be made available. With static type checking, programmers are relieved from the task of checking manually if the correct types are used.

Although type checking is so important for C++ practitioners, the need of naming types at each variable and function declaration has become too burdensome for some programmers. After all, every expression in C++ has a type, and with the introduction of containers and other templates, it becomes sometimes difficult to write the proper type of an expression. To avoid this problem, the C++ committee decided to use the auto keyword to allow for automatic type detection in C++ expressions. This feature was fist introduced in C++11 but has been progressively extended through each standard until C++20.

Automatic type deduction frees programmers from the need to indicate the type of each variable when declaring it. The type deduction system works through the use of information that is already available to the compiler at the moment an expression is being parsed. For example, if a variable is created from a known constant, the compiler can easily determine its type. On the other hand, if a variable is initialized to the result of an expression, the compiler can also determine the type of the result and use it for the variable. Here are some simple examples:

```
void autoExample()
{
    auto i = 1;              // this is an integer
    auto d = 2.0;            // this is a float
    auto d2 = d + 1;         // this is also a float
    auto str = "hello";      // this is a char *

    cout << "integer : " << i << " float: "
        << d2               << " string " << str << endl;
}
```

Here, the first, second, and fourth variables are initialized using constants, so the type is immediate. The third variable has its type determined through the result of the expression given as initializer.

Another area where auto variables are very useful is when working with templates. Many templates generate complex types, which are difficult to type and to remember. It is very useful to be able to avoid typing these types with the help of the auto keyword. Here is an example using an iterator to an STL container:

```
void autoTemporaryExample()
{
    std::vector<std::pair<int,std::string>> myVector;

    // without auto
    for (std::vector< std::pair<int,std::string>>::iterator
        it = myVector.begin();
        it != myVector.end(); ++it)
    {
        // do something here
    }

    // with auto
    for (auto it = myVector.begin(); it != myVector.end(); ++it)
    {
        // same thing here
    }
}
```

The first loop shows the type of the iterator used to visit all members of the container. It is even harder to type than the original template name. The second loop shows how to express the same thing using the auto keyword. Here, it is possible to avoid the name of the template, which makes it much easier to understand what the code is doing.

Another way in which the auto keyword is used is to determine parameter types for template functions. This is a more recent use of auto, added in the C++20 standard, but it follows the same pattern: the type of the parameter is determined by the compiler as it determines this information from the actual parameters. Here is an example:

```
auto add_args(auto x, auto y) {
    return x + y;
}

int main() {
    int a = 10;
    double b = 20;
    auto res = add_args(a, b);
    return 0;
}
```

Notice that without the help of the auto operator, this would be declared in the following way:

```
template <class A, class B>
A add_args2(A x, B y)  {
    return x + y;
}

int main() {
    int a = 10;
    double b = 20;
    int  res = add_args2(a,b);
    return 0;
}
```

Lambdas

A lambda is a function that can be created on the spot, without the need for a separate top-level declaration. Lambda functions can, additionally, be allowed to retain references to variables that exist at the same level in which they are introduced. A lambda function can be saved in variables and passed to other functions, where they can be used as needed. The variables that have been saved in the context can be used even after the original block has finished. Here is a very simple example:

```
void lambdaExample()
{
    auto avg = [](int a, int b) { return (a + b) / 2; };

    cout << "the average of 3 and 5 is "  << avg(3, 5) << endl;
}
```

The syntax for lambda functions starts with an angle bracket. The return type doesn't need to be specified, and it is deduced from the variable or expression in the return keyword. Here is an example where there is local variable capture:

```
void lambdaExample2()
{
    double factor = 2.5;
    auto scaledAvg = [&factor](int a, int b) {
        return factor * (a + b) / 2;
    };

    auto modifiedAvg = [&](int a, int b) { return scaledAvg(a, b); };

    cout << "the scaled average of 3 and 5 is "
        << scaledAvg(3, 5) << endl;
    cout << "this should be the same "
        << modifiedAvg(3, 5) << endl;
}
```

The example shows two lambda functions where there is variable capture. In the first function, the factor variable is captured and becomes available to be used inside the lambda function. The second example shows a lambda function where all local variables are captured (indicated by the [&] notation). In this case, any local variable can be used, including the scaledAvg variable.

User-Defined Literals

You are familiar with literals for standard types such as int, float, or char. These literal values allow one to initialize new variables as needed. C++11 introduces user-defined literals, where a literal can be manipulated to perform any kind of preprocessing. This is useful in the case that scalar numbers need to go though some kind of conversion before they are used as initializers.

The syntax used for user-defined literals is similar to other operators. The operator " " keyword is used to introduce the new literal format. Consider an example where you wish to define numeric literals that return the price in Euros. This can be defined in the following way:

```
long double operator "" _eu(long double val)
{
    return val / 1.24;
}
```

Notice the signature that contains the name operator " ", followed by the suffix _eu. In this case, you'll be using a fixed conversion value, but in general, you could have a more complex scheme for conversion from dollars to euros. Finally, you can use this user-defined literal in the following way:

```
void showUserDefinedLiterals()
{
    double price = 300; // price in dollars
    long double priceInEU =  300.0_eu;

    cout << " price in dollars: "  << price
        << " price in Euros: " << priceInEU << endl;
}
```

Here, you first define a price without any conversion (in dollars). Then you create a second variable that corresponds to the same quantity, but using the user-defined suffix _eu. Using this suffix, you will have a converted price in the priceInEU variable, as printed at the end of the showUserDefinedLiterals function.

Range-Based for

STL containers are some of the most used templates in any C++ system. These containers are versatile and can be used to perform a large nuber of operations to its components. In the previous versions of C++, it was possible to iterate through the components of a container using an auxiliary iterator variable. For example:

```
void loopExample1()
{
    std::vector<std::pair<double,std::string>> v;

    // without auto
    for (std::vector<std::pair<double,std::string>>::iterator it =
    v.begin();
        it != v.end(); ++it)
    {
        // do something here
    }
}
```

Or you can use an auto variable to simplify the code above a little. Still, there is a lot of code necessary just to iterate over the elements of the container. The C++11 standard introduces a simpler way to do this, with the container-oriented for loop. The syntax for this special case is simplified, so you don't need to write the boundary conditions (begin() and end()) for the container. Here is the preceding example, modified to use the new for loop.

```
void forLoopExample()
{
    std::vector<std::pair<double,std::string>> vectorOfPairs;

    for (auto &i : vectorOfPairs)
    {
        cout << " values are "
            << i.first << " and "
            << i.second << endl;
    }
}
```

Notice how the vectorOfPairs variable is now used only once in the second part of the loop statement. The auto variable declaration avoids the need for a long template declaration, which helps to keep the notation easy to read.

Rvalue References

One of the common performance issues with the use of containers and strings in C++ is the fact that temporary variables need to be created in so many places:

- When moving elements between two containers, it is frequently necessary to perform a copy and then delete the old elements.

- When implementing operators, it is often necessary to return new objects each time an operation is performed, since the argument to an operator (such as <<) may very well be a temporary object.

- When returning objects from functions, it becomes necessary to copy the return object to a temporary, since it belonged to a function that is finishing. If this temporary object is immediately assigned to a new variable, then the temporary object is not used.

To help developers tackle these issues, C++ designers decided to introduce a notation for variables that are not named and that cannot be assigned outside of the current context. Such variables are known as rvalues, because in any expression, they can only appear in the right side of the assignment. Examples of rvalues are immediate values passed as parameters to functions and temporaries created during the evaluation of expressions, among others.

The syntax for rvalues is similar to references, but with the && sign used instead of a single & sign. Such declarations are useful mainly in the list of arguments for a function, as well as in the return. Here are some examples of their use:

```
#include <string>
using std::string

void rvalExamp(string &&s)
{
    cout << " string is " << s << endl;
}
```

```
void rvalExamp(string &s)
{
    cout << " string lvalue: " << s << endl;
}

int main()
{
    rvalExamp("a test string");  // calls rval version
    string a = "string a ";
    string b = "string b ";
    rvalExamp(a + b);            // calls rval version
    string c = "another example";
    rvalExamp(c);               // calls lval version
    return 0;
}
```

In this example, any string (including temporary values) can be passed to the function rvalExamp. The rvalue may be used with the knowledge that its temporary value will be destroyed at the end of the function. On the other hand, you can also have a version of the function that receives a standard lvalue reference. This version of the function is called only when a lvalue is used as parameter (in this case, it happens when the parameter is a named variable).

An important case where rvalues may be useful is in the assignment operator. If the parameter to the operator is a rvalue, then it is usually possible to optimize it by reducing the number of allocations. This is shown in the following example:

```
#include <vector>
using std::vector;

class RValTest {
public:
    RValTest(int n);
    RValTest(const RValTest &x);
    ~RValTest();

    RValTest &operator=(RValTest &&p);  // this is for RVAL
    RValTest &operator=(RValTest &p);   // this is for LVAL
```

```cpp
private:
    vector<int> data;
};

RValTest::RValTest(int n)
: data(n, 0)
{

}

RValTest::RValTest(const RValTest &p)
: data(p.data)
{
}

RValTest::~RValTest()
{
}

RValTest &RValTest::operator=(RValTest &&p)
{
    data.swap(p.data);
    cout << " calling rval assignment " << endl;
    return *this;
}

RValTest &RValTest::operator=(RValTest &p)
{
    if (this != &p)
    {
        data = p.data;
    }
    cout << " calling normal assignment " << endl;
    return *this;
}
```

```
void useRValTest()
{
    RValTest test(3);
    RValTest test2(4);

    test2 = test;          // use standard assignment
    test = RValTest(5);   // use rval assignment
}
```

The class RValTest knows when the assignment operator is called with a temporary. In this case, you can just swap the elements of the data array, instead of performing expensive data copy.

New Function Declarator Syntax and decltype

You have seen that the keyword auto was repurposed to allow for automatic type deduction or variables. However, once this change has been made to how variables are declared, soon you will also need to return such values. For example, consider the following function:

```
void autoFunctExample1(vector<int> &x)
{
    auto iterator = x.begin();
    // do something with iterator
}
```

This works fine, and you don't need to know the exact template type returned by begin() to use it. However, a big problem arises if you need to return the variable iterator. In that case, you need to somehow determine the type of iterator just to declare the function, since the return type must be part of the signature.

To help solve this problem, C++11 introduced a new form of function declaration, which uses auto instead of the name of the type. Still, do maintain the type checking system the compiler needs to determine the type of a function. This is where the decltype keyword comes in. The decltype operator returns the type of any expression that is given as a parameter. Similarly to how sizeof returns information from a type, decltype returns the type for a variable or other general expression.

Using decltype, you can now add a return type declaration to a function after the ->
operator, which may only appear right after the list of arguments to the function. Since at
this point the type of the arguments to the function are known, you can use them along
with decltype to define the return type. Here is an example based on the code presented
previously:

```
auto autoFuncExample(vector<int> &x) -> decltype(x.begin())
{
    auto iterator = x.begin();
    // do something with iterator
    return iterator;
}
```

Now you can return the iterator without knowing its exact type, since it is
automatically calculated during compilation time.

The decltype operator is not restricted to appear in the declaration of a return type.
You can use it anywhere a type may be required, although many times, it can be
substituted by the auto keyword. For example, the variable declaration auto x = 1
is equivalent to decltype(1) x = 1. But decltype can be used in other contexts,
such as sizeof(decltype(x.begin())), to determine the size of a deduced
type, where auto would not work.

Delegating Constructors

In older versions of C++, the problem of creating and maintaining initializers along with
constructors was well known. For example, you needed to initialize all scalar variables
in the same order that it appears in the class declaration. C++11 avoids this issue by
delegating the task of data initializing to other constructors.

A delegating constructor is simply one that can be used by other constructors,
so to avoid the repetition of data initialization statements. For example, suppose you
have a class Dimensions with three member variables. You can have three different
constructors, each one accepting a different number of components for this dimension

object. To avoid repeating yourself during the initialization part, you can create a single initializer constructor and call this constructor from the others. Here is a possible implementation using C++11:

```cpp
class Dimensions {
public:
    Dimensions();
    Dimensions(double x);
    Dimensions(double x, double y);
    Dimensions(double x, double y, double z);

private:
    double m_x;
    double m_y;
    double m_z;
};

Dimensions::Dimensions()
: Dimensions(0, 0, 0)
{
}

Dimensions::Dimensions(double x)
: Dimensions(x, 0, 0)
{
}

Dimensions::Dimensions(double x, double y)
: Dimensions(x, y, 0)
{
}

Dimensions::Dimensions(double x, double y, double z)
: m_x(x),
  m_y(y),
  m_z(z)
{
}
```

The constructor Dimensions (double x, double y, double z) is the only one that can access the member variables directly, while the others are only using it to perform indirect initialization.

Inheriting Constructors

Another common problem in earlier versions of C++ was the handling of constructors in derived classes. Sometimes, a constructor derived from a class has constructors that are identical to the constructors in the superclass. In this case, it was necessary to replicate all constructors in the subclass so that it would become available to clients. It seems clear that this is an undesirable code replication, and it was addressed by the C++11 standard. Now, it is possible to employ the using keyword to introduce the constructors existing in the base class. Here is an example, using the Dimensions class as its base.

```
class DimensionsDerived : public Dimensions {
public:
    using Dimensions::Dimensions;

};

int main()
{
    DimensionsDerived(1, 2, 4);
}
```

The new class can be created using the same constructors as the parent, since it contains the using declaration for the base constructor.

Generalized Attributes

Attributes provide a standard syntax for the addition of annotations to elements contained in C++ code. Most compilers use nonstandard mechanisms to determine the attributes of certain elements. For example, if a function can be exported or not is defined by attributes, which varied for each compiler vendor.

C++14 introduced a new syntax for attributes that can be used by any compiler vendor. The attributes are introduced inside double brackets and contain annotation for the element that is syntactically next to the attribute. Here is an example:

```
struct [[exported]] AttribSample
{
    int memberA;
    [[gnu::aligned (16)]]
    double memberB;
};
```

Note   The list of available attributes is specific to each compiler. However, at least for gcc, it is possible to write custom plugins that are able to process these attributes. For example, suppose that you create a plug-in to process GUI-based classes in your code base. Running gcc with that plug-in will let you perform actions for each GUI class, such as generating additional code, creating resources files to be used during run time, and other related tasks.

Generalized Constant Expressions

In modern C++, we have a great emphasis on the use of templates and related compile-time programming techniques. The STL and many other well-known libraries, such as boost, depend heavily on templates. However, since template-based operations are compile-time by definition, they introduce the need for constant, compile-time evaluated expressions. Such expressions have in common the fact that they evaluate to constant values, so that all the results will be available at compilation time.

While normal C++ code can involve both runtime and compilation-time expressions, it is useful to guarantee that the value in a particular function is completely evaluated at compilation time. This cannot be guaranteed with traditional function, however, which motivated the standards committee to introduce constant expressions as a compiler-enforced concept.

To guarantee that a function will evaluate only to constants that are available at compiler time, you should use the new `constexpr` keyword. When this keyword is added before a function declaration, the compiler will force its evaluation and emit an error if the included expressions cannot be calculated at compile time. Here is a simple example:

```
struct TestStruct {
    int a;
    char b;
    double c;
};

template <class T>
constexpr int testDataSize(T)
{
    return sizeof(T);
}

constexpr int minTestSize()
{
    return 2 * testDataSize(TestStruct()) + 1;
}
```

The `testDataSize` function just shows how easy it is to create a compile-time function. The return value calculated in the first function is the size of a test data structure, which can later be used by other constant calculations. The second function just calculates what is considered the minimum size for the test data in the application. Results such as the ones presented previously can be freely used on templates, as a way to perform more complex calculations.

Null Pointer Constant

A null pointer is a pointer that doesn't correspond to any valid address in the target machine. Traditionally, null pointer values have been used to indicate that a pointer is not in use. For functions returning values, this usually means that the desired pointer is invalid, among other possible uses.

C++ inherited from C the idea that null pointers are equivalent to the constant zero, since this is an invalid pointer value in most computer architectures. In fact, the preprocessor macro NULL is defined in previous versions of C and C++ as 0. The fact that the 0 value can be confused with NULL in a numeric context, however, is one of the problems inherent to this definition. To simplify the rules concerning null pointers, the C++ committee decided to introduce a new keyword, nullptr, which can only be interpreted as a pointer and not an integer or any other type that is related to the 0 constant.

```
void *testNull()
{
    int *pi = new int;
    if (pi == nullptr)
    {
        return nullptr;
    }
    // *pi = 1 + nullptr; // this doesn't work, nullptr is not an integer
    return pi;
}
```

The preceding code checks if a newly allocated variable is null. Notice that the nullptr value cannot be used to simultaneously initialize an integer variable: it can only be used in a pointer context.

Defaulted and Deleted Member Functions

Another new feature in C++ is the ability to clearly determine if a class will use or disallow any of the default member functions provided by the compiler. Remember that there are four member functions automatically provided when a class is created:

- The default constructor

- The copy constructor

- The move copy constructor (because of new move semantics)

- The destructor

- The assignment operator

- The move assignment operator (because of new move semantics)

Standard practice indicates that you should define these functions for every new type, as you can see in the examples presented in this book. However, C++ gives another option: you can use the `default` and `delete` keywords to determine which of these member functions can be used by default (with the version created by the compiler) and which versions should be discarded. For example:

```cpp
class TestDefaults {
public:
    TestDefaults() = default;
    TestDefaults(int arg);
    TestDefaults(const TestDefaults &) = delete;

    // other member functions here
};
```

This class uses a default constructor, whose definition is written automatically by the compiler, even though it has a non-default constructor that receives a single integer argument. This was not possible in previous versions of C++, where you could either accept the default constructor or write it again in case you wanted two or more constructors. Notice that you can, at the same time, reject the default copy constructor. Therefore, the previous declaration directly indicates that the type cannot be copied.

Another useful feature of default member functions is that you can introduce virtual destructors without the need to write one. Remember that classes that include virtual member functions also require virtual destructors in order to clean up resources in each of the levels of the class hierarchy. The standard way of doing this is introducing an empty virtual destructor, in order to allow for virtual destructors in the subclasses. In C++11, you can use the default keyword to provide a default, virtual destructor. In the previous example, this could be added in the following way:

```cpp
class TestDefaults {
public:
    TestDefaults() = default;
    TestDefaults(int arg);
    virtual ~TestDefaults() = default;
    TestDefaults(const TestDefaults &) = delete;

    // other member functions here
};
```

Notice that you don't need to explicitly write the destructor, since it will use the default implementation. The derived classes, however, will enjoy the use of a virtual destructor due to the definition in the base class.

Initializer Lists

One of the confusing aspects of C++ syntax is initialization. Different objects, such as integers, structures, classes, and arrays, have slightly different ways to be initialized. C++11, while maintaining the previous methods for variable initialization, introduces a new way to perform initialization that is much more regular and can be applied to any object in the language.

The syntax for initialization lists uses braces to surround one or more constants or variables. These elements are then applied to the new variable and interpreted according to its type. Here are a few examples:

```cpp
void initializationTest()
{
    int x {}; // equivalent to int x = 0;
    int y { 0 };  // same as above
    const char *s { "var"  };
    double d { 2.4 };
    struct StrTest {
        int a;
        double d;
        char c;
    };

    StrTest structVal { 2, 4.2, 'f' };

    cout << " values are "  << x << "  " << y << " "  << s
        << " "  << d << " " << structVal.a << endl;
```

```
class AClass {
public:
    AClass(int v) : m_val(v) {}
    int m_val;
};

AClass obj = { 3 } ;
}
```

Notice that all these values can be easily initialized using the brace notation. Among the advantages of this strategy is the fact that you can also initialize containers (such as vectors) using lists of values enclosed in braces. Here is an example of this feature:

```
#include <map>
#include <vector>
Using std::vector;
using std::map;

void containerInitialization()
{
    vector<int> vi = { 1, 3, 5, 7, 9, 11 };

    for (auto &v : vi)
    {
        cout << v << " ";
    }

    map<int,double> m = { { 2, 3.0}, {4, 5.0} };
    for (auto &v : m)
    {
        cout << v.first << " " << v.second << " ";
    }
}
```

You can see from the previous example how initialization lists can be effectively used to pass data to standard containers found in the STL. Most containers in C++11 have one or more constructors that can receive initialization lists. Finally, you can also create classes that receive lists of parameters, using the class std::initializer_list. The compiler will automatically fill the initializer_list container with the values passed to the constructor.

```cpp
class MyClass {
public:
    MyClass(std::initializer_list<int> args);
    vector<int> m_vector;
};

MyClass::MyClass(std::initializer_list<int> args)
{
    m_vector.insert(m_vector.begin(), args.begin(), args.end());
}

void useClassInitializer()
{
    MyClass myClass = { 2, 5, 6, 22, 34, 25 };

    for (auto &v : myClass.m_vector)
    {
        cout << v << " ";
    }
}
```

Index

© Carlos Oliveira 2021
C. Oliveira, *Practical C++20 Financial Programming*, https://doi.org/10.1007/978-1-4842-6834-6

Printed in the United States
by Baker & Taylor Publisher Services